Linda Meeks
The Ohio State University

Philip Heit
The Ohio State University

McGraw Hill Education

Front Cover: McGraw-Hill Education

Title Page: McGraw-Hill Education

Back Cover: McGraw-Hill Education

MHEonline.com

Send inquiries to:
McGraw-Hill Education
8787 Orion Place
Columbus, OH 43240

ISBN: 978-0-07-667504-3
MHID: 0-07-667504-1

Printed in the United States of America.

6 7 8 9 10 QVS 21 20 19 18 17

Contributors

Celan Alo, M.D., MPH
Medical Epidemiologist
Bureau of Chronic Disease and Tobacco Prevention
Texas Department of Health
Austin, Texas

Danny Ballard, Ed.D.
Associate Professor, Health
Texas A&M University
College of Education
College Station, Texas

Lucille Villegas Barrera, M.Ed.
Elementary Science Specialist
Houston Independent School District
Houston, Texas

Gus T. Dalis, Ed.D.
Consultant of Health Education
Torrance, California

Alisa Evans-Debnam, MPH
Dean of Health Programs
Fayetteville Technical Community College
Fayetteville, North Carolina

Susan C. Giarratano-Russell, MSPH, Ed.D., CHES
Health Education, Evaluation & Media Consultant
National Center for Chronic Disease Prevention & Health Promotion
Centers for Disease Control & Prevention
Glendale, California

Donna Lloyd-Kolkin, Ph.D.
Principal Associate
Public Health Applications & Research
Abt Associates, Inc.
Bethesda, Maryland

Mulugheta Teferi, M.A.
Principal
Gateway Middle School
Center for Math, Science & Technology
St. Louis, Missouri

Roberto P. Treviño, M.D.
Director, Social & Health Research Center
Bienestar School-Based Diabetes Prevention Program
San Antonio, Texas

Dinah Zike, M.Ed.
Dinah Might Adventures LP
San Antonio, Texas

Content Reviewers

Mark Anderson
Supervisor, Health Physical Education
Cobb County Public Schools
Marietta, Georgia

Ken Ascoli
Assistant Principal
Our Lady of Fatima High School
Warren, Rhode Island

Jane Beougher, Ph.D.
Professor Emeritus of Health Education, Physical Education, and Education
Capital University
Westerville, Ohio

Lillie Burns
HIV/AIDS Prevention Education
Education Program Coordinator
Louisiana Department of Education
Baton Rouge, Louisiana

Jill English, Ph.D., CHES
Professor, Soka University
Aliso Viejo, California

Elizabeth Gallun, M.A.
Specialist, Comprehensive Health Education
Maryland State Department of Education
Baltimore, Maryland

Brenda Garza
Health Communications Specialist
Centers for Disease Control & Prevention
Atlanta, Georgia

Sheryl Gotts, M.S.
Consultant, Retired from Milwaukee Schools
Milwaukee, Wisconsin

Russell Henke, M.Ed.
Coordinator of Health
Montgomery County Public Schools
Rockville, Maryland

Kathy Kent
Health and Physical Education Teacher
Simpsonville Elementary School at Morton Place
Simpsonville, South Carolina

Bill Moser, M.S.
Program Specialist for Health and Character Education
Winston-Salem Forsyth City Schools
Winston-Salem, North Carolina

Debra Ogden
Curriculum Coordinator
District School Board of Collier County
Naples, Florida

Thurman Robins
Chair/Professor
Health and Kinesiology Department
Texas Southern University
Houston, Texas

Sherman Sowby, Ph.D., CHES
Professor, Department of Health Science
California State University, Fresno
Fresno, California

Greg Stockton
Health and Safety Expert
American Red Cross
Washington, D.C.

Deitra Wengert, Ph.D., CHES
Professor, Department of Health Science
Towson University
Towson, Maryland

Susan Wooley-Goekler, Ph.D., CHES
Adjunct Faculty
Kent State University
Kent, Ohio

Medical Reviewers

Celan Alo, M.D., MPH
Medical Epidemiologist
Bureau of Chronic Disease and Tobacco Prevention
Texas Department of Health
Austin, Texas

Donna Bacchi, M.D., MPH
Associate Professor of Pediatrics
Director, Division of Community Pediatrics
Texas Tech University Health Science Center
Lubbock, Texas

Olga Dominguez Satterwhite, R.D., L.D.
Registered Dietitian and Diabetes Educator
Baylor College of Medicine
Houston, Texas

Roberto P. Treviño, M.D.
Director, Social & Health Research Center
Bienestar School-Based Diabetes Prevention Program
San Antonio, Texas

Contents

UNIT A Mental, Emotional, Family, and Social Health

CHAPTER 1 Mental and Emotional Health

LESSON 1 What Are Health and Wellness? **A4**
CONSUMER WISE: Top Ten Risk Behaviors in TV Shows A7
Write About It! A8
LIFE SKILLS ACTIVITY: Practice Healthful Behaviors A9

LESSON 2 Plan for a Healthy Life **A10**
ON YOUR OWN: Make Health Facts Yellow Pages A11
BUILD CHARACTER: Plan to Improve A12
Write About It! A13
LIFE SKILLS ACTIVITY: Set Health Goals A15

LESSON 3 Your Personality and Character **A16**
LINK Art A17
Write About It! A19
MAKE A DIFFERENCE: Care Bags A20
LIFE SKILLS ACTIVITY: Be a Health Advocate A21

LESSON 4 Your Emotions **A22**
CAREERS: Psychologist A23
LINK Art A24, A27
LIFE SKILLS ACTIVITY: Use Communication Skills A24
HEALTH ONLINE: Emotions and Health A25
Write About It! A26

LESSON 5 Taking Charge of Your Health **A28**
Write About It! A29
CONSUMER WISE: Spot the Sales Pitch A30
BUILD CHARACTER: Help Resist Pressure A32
LIFE SKILLS ACTIVITY: Use Resistance Skills A33

LESSON 6 Managing Stress **A34**
ON YOUR OWN: Show Your Skills A37
Write About It! A38
LIFE SKILLS ACTIVITY: Manage Stress A39
LEARNING LIFE SKILLS: Make Responsible Decisions A40

CHAPTER REVIEW A42

CHAPTER 2 Family and Social Health

LESSON 1 Your Social Health **A46**
CONSUMER WISE: Analyze Relationships in Ads A47
Write About It! A48
LINK Physical Education A49
LINK Art A50
LIFE SKILLS ACTIVITY: Be a Health Advocate A51

LESSON 2 Communication in Relationships **A52**
Write About It! A54
LINK Physical Education A56
LIFE SKILLS ACTIVITY: Use Communication Skills A57

LESSON 3 When Conflict Occurs **A58**
ON YOUR OWN: Sorry Time A60
Write About It! A62
LIFE SKILLS ACTIVITY: Resolve Conflicts A63

LESSON 4 Health in the Family **A64**
BUILD CHARACTER: Private Journal A65
LINK Science A66
LIFE SKILLS ACTIVITY: Analyze What Influences Your Health A67

LESSON 5 Facing Family Challenges **A68**
LINK Art A69
BUILD CHARACTER: Role-Play an Adoption A70
CAREERS: Sociologist A71
Write About It! A72
LIFE SKILLS ACTIVITY: Access Health Facts A73

LESSON 6 Among Friends **A74**
LINK Music A76
LIFE SKILLS ACTIVITY: Set Health Goals A77
HEALTH ONLINE: Resolving Conflicts A78

LESSON 7 Facing Challenges in Relationships **A80**
ON YOUR OWN: Interview a Family Member A82
LIFE SKILLS ACTIVITY: Make Responsible Decisions A83
LEARNING LIFE SKILLS: Use Resistance Skills A84

CHAPTER REVIEW A86

UNIT A ACTIVITIES AND PROJECTS A88

UNIT B **Growth and Nutrition**

CHAPTER 3 Growth and Development

LESSON 1 Your Body's Systems **B4**

LINK Science B5

LINK Physical Education B7

CAREERS: Nurse Practitioner B8

LIFE SKILLS ACTIVITY: Manage Stress B9

LESSON 2 Your Heart and Lungs **B10**

Write About It! B12

LIFE SKILLS ACTIVITY: Be a Health Advocate B13

LESSON 3 More Body Systems **B14**

CONSUMER WISE: Analyze Ads B15

HEALTH ONLINE: System Savvy B16

LIFE SKILLS ACTIVITY: Access Health Products B17

ON YOUR OWN: Drink More Water B19

LESSON 4 The Stages of Life **B20**

LINK Social Studies B23

BUILD CHARACTER: Make an Album B24

LIFE SKILLS ACTIVITY: Make Responsible Decisions B25

LESSON 5 You Are Unique **B26**

Write About It! B27

LINK Art B28

LIFE SKILLS ACTIVITY: Analyze What Influences Your Health B29

LEARNING LIFE SKILLS: Practice Healthful Behaviors B30

CHAPTER REVIEW B32

CHAPTER 4 Nutrition

LESSON 1 Your Basic Nutritional Needs **B36**

LINK Art B38

LIFE SKILLS ACTIVITY: Access Health Facts B39

CONSUMER WISE: Read the Labels B41

LESSON 2 Aim for a Balanced Diet **B42**

LINK Social Studies B43

ON YOUR OWN: Plan a Menu B44

BUILD CHARACTER: Plan to Share Food B45

LIFE SKILLS ACTIVITY: Use Resistance Skills B47

LEARNING LIFE SKILLS: Analyze What Influences Your Health B48

LESSON 3 Food That's Safe to Eat **B50**

CAREERS: Food Inspector B51

HEALTH ONLINE: Keeping It Safe B52

MAKE A DIFFERENCE: Ryan's Wells B53

LINK Science B54

LIFE SKILLS ACTIVITY: Practice Healthful Behaviors B55

LESSON 4 Your Weight Manager **B56**

LINK Math B57

Write About It! B58

LINK Art B59

LIFE SKILLS ACTIVITY: Make Responsible Decisions B60

CHAPTER REVIEW B62

UNIT B ACTIVITIES AND PROJECTS B64

UNIT C

Personal Health and Safety

CHAPTER 5 Personal Health and Physical Activity

LESSON 1 Caring for Your Body **C4**

LIFE SKILLS ACTIVITY: Access Health Facts C5

ON YOUR OWN: See What Skin Does C6

LESSON 2 Your Teeth, Eyes, and Ears **C10**

ON YOUR OWN: Avoid Eyestrain C12

LINK Science C14

LIFE SKILLS ACTIVITY: Make Responsible Decisions C14

LESSON 3 The Benefits of Physical Activity **C16**

Write About It! C17

LIFE SKILLS ACTIVITY: Practice Healthful Behaviors C21

LESSON 4 A Balanced Workout **C22**

CONSUMER WISE: Make a Buying Guide C25

LINK Math C26

LIFE SKILLS ACTIVITY: Set Health Goals C29

LESSON 5 Play It Safe **C30**

HEALTH ONLINE: Gear Up for Safety C31

LINK Physical Education C32

CAREERS: Personal Trainer C33

BUILD CHARACTER: Perform a Skit C34

LIFE SKILLS ACTIVITY: Use Communications Skills C35

LEARNING LIFE SKILLS: Access Health Products C36

CHAPTER REVIEW C38

CHAPTER 6 Violence and Injury Prevention

LESSON 1 Keep Safe Indoors **C42**

LINK Science C43

CAREERS: Firefighter C44

ON YOUR OWN: Make an Escape Plan C45

LIFE SKILLS ACTIVITY: Set Health Goals C47

LESSON 2 Keep Safe Outdoors **C48**

MAKE A DIFFERENCE: Safer Streets C49

Write About It! C50

LIFE SKILLS ACTIVITY: Be a Health Advocate C51

LESSON 3 How to Handle Emergencies **C54**

ON YOUR OWN: Role-Play C55

HEALTH ONLINE: Get Weather Ready C56

LIFE SKILLS ACTIVITY: Access Health Facts C59

LESSON 4 Facts on First Aid **C60**

LINK Science C61

Write About It! C63

LIFE SKILLS ACTIVITY: Make Responsible Decisions C65

LESSON 5 Staying Violence Free **C66**

CONSUMER WISE: Violence on TV C67

LIFE SKILLS ACTIVITY: Use Communication Skills C69

LINK Social Studies C70

LEARNING LIFE SKILLS: Resolve Conflicts C72

LESSON 6 Steering Clear of Gangs **C74**

BUILD CHARACTER: Include Others C75

LINK Music C76

LIFE SKILLS ACTIVITY: Use Resistance Skills C77

CHAPTER REVIEW C78

UNIT C ACTIVITIES AND PROJECTS C80

UNIT D Drugs and Disease Prevention

CHAPTER 7 Alcohol, Tobacco, and Other Drugs

LESSON 1 Drugs and Your Health **D4**

CONSUMER WISE: Design a Brochure D5

ON YOUR OWN: Find Expiration Dates D7

Write About It! D8

LIFE SKILLS ACTIVITY: Make Responsible Decisions D9

LESSON 2 Alcohol and Health **D10**

Write About It! D12

LINK Science D13

LIFE SKILLS ACTIVITY: Set Health Goals D15

LESSON 3 Tobacco and Health **D16**

LINK Art D17

LINK Science D18

LINK Math D19

LINK Music D20

LIFE SKILLS ACTIVITY: Analyze What Influences Your Health D21

LEARNING LIFE SKILLS: Be a Health Advocate D22

LESSON 4 Other Drugs to Avoid **D24**

Write About It! D25

CAREERS: Drug Counselor D27

LIFE SKILLS ACTIVITY: Manage Stress D29

LESSON 5 When Someone Abuses Drugs **D30**

ON YOUR OWN: Fight Pressure D31

Write About It! D33

BUILD CHARACTER: List Reasons to Respect Yourself D34

LIFE SKILLS ACTIVITY: Use Communication Skills D35

LESSON 6 Resisting Pressure **D36**

HEALTH ONLINE: Explore Government Agencies D38

LIFE SKILLS ACTIVITY: Use Resistance Skills D39

CHAPTER REVIEW D40

CHAPTER 8 Communicable and Chronic Diseases

LESSON 1 Communicable Diseases **D44**

BUILD CHARACTER: Plan to Prevent Disease D47

LINK Art D48

LIFE SKILLS ACTIVITY: Practice Healthful Behaviors D49

LESSON 2 How Your Body Fights Infection **D50**

CAREERS: Public Health Worker D52

ON YOUR OWN: Record Your Vaccines D53

Write About It! D54

LIFE SKILLS ACTIVITY: Access Health Services D55

LESSON 3 Signs of Illness **D56**

LIFE SKILLS ACTIVITY: Make Responsible Decisions D59

LESSON 4 Chronic Disease and the Heart **D60**

LINK Science D61

LINK Science D62

CONSUMER WISE: Check for "Heart-Smart" Foods D64

LIFE SKILLS ACTIVITY: Set Health Goals D65

LESSON 5 Chronic Disease: Cancer **D66**

LINK Math D67

MAKE A DIFFERENCE: Gifts for Families. D68

LIFE SKILLS ACTIVITY: Be a Health Advocate D69

LEARNING LIFE SKILLS: Manage Stress D70

LESSON 6 Other Chronic Diseases **D72**

HEALTH ONLINE: Explore Chronic Diseases D73

Write About It! D74

CONSUMER WISE: Write a Commercial D75

LINK Science D76

LIFE SKILLS ACTIVITY: Analyze What Influences Your Health. D77

CHAPTER REVIEW D78

UNIT D ACTIVITIES AND PROJECTS D80

UNIT E

Community and Environmental Health

CHAPTER 9 Consumer and Community Health

LESSON 1 What Smart Consumers Know **E4**

CAREERS: Consumer Reporter E5

Write About It! E6

LIFE SKILLS ACTIVITY: Make Responsible Decisions E7

LINK Science E8

LESSON 2 Help for Consumers **E10**

CONSUMER WISE: Design a Brochure E11

HEALTH ONLINE: Budget Benefits E12

LIFE SKILLS ACTIVITY: Access Health Facts E13

LEARNING LIFE SKILLS: Use Communication Skills E14

LESSON 3 Planning a Healthful Community **E16**

LINK Social Studies E17

LINK Art E18

ON YOUR OWN: Identify Recycled Products E19

BUILD CHARACTER: Clean Up a Park E20

LIFE SKILLS ACTIVITY: Analyze What Influences Your Health E21

LESSON 4 Careers in Health and Fitness **E22**

Write About It! E23

ON YOUR OWN: Design a Dream Job E25

ON YOUR OWN: Diagram a Career E26

LIFE SKILLS ACTIVITY: Set Health Goals E27

CHAPTER REVIEW E28

CHAPTER 10 Environmental Health

LESSON 1 Your Environment **E32**
Write About It! E33
LINK Science E34
LIFE SKILLS ACTIVITY: Use Communication Skills E35
LINK Social Studies E36
LEARNING LIFE SKILLS: Set Health Goals E38

LESSON 2 Protecting Water and Land **E40**
CONSUMER WISE: Design a Safety Poster E41
HEALTH ONLINE: Watch the Waste E42
Write About It! E43
LIFE SKILLS ACTIVITY: Make Responsible Decisions E43
CAREERS: Recycling Plant Worker E44
ON YOUR OWN: Dispose of Hazardous Wastes E45

LESSON 3 Conservation **E46**
LINK Math E47
LINK Art E48
LIFE SKILLS ACTIVITY: Practice Healthful Behaviors E49

LESSON 4 A Positive Environment **E50**
MAKE A DIFFERENCE: Reducing Trash E51
BUILD CHARACTER: Volunteer to Help E52
LIFE SKILLS ACTIVITY: Be a Health Advocate E53

CHAPTER REVIEW E54

UNIT E ACTIVITIES AND PROJECTS E56

Features and Activities

UNIT A Mental, Emotional, Family, and Social Health

Learning Life Skills
Make Responsible Decisions, **A40**
Use Resistance Skills, **A84**

Life Skills Activities
Practice Healthful Behaviors, **A9**
Set Health Goals, **A15, A77**
Be a Health Advocate, **A21, A51**
Use Communication Skills, **A24, A57**
Use Resistance Skills, **A33**
Manage Stress, **A39**
Resolve Conflicts, **A63**
Analyze What Influences Your Health, **A67**
Access Health Facts, **A73**
Make Responsible Decisions, **A83**

Build Character
Plan to Improve, **A12**
Help Resist Pressure, **A32**
Private Journal, **A65**
Role-Play an Adoption, **A70**

Cross Curricular Links
Art, **A17, A24, A27, A50, A69**
Physical Education, **A49, A56**
Science, **A66**
Music, **A76**
Write About It!, **A8, A13, A19, A20, A26, A29, A38, A48, A54, A62, A72**

On Your Own for School or Home
Make Health Facts Yellow Pages, **A11**
Show Your Skills, **A37**
Sorry Time, **A60**
Interview a Family Member, **A82**

Consumer Wise
Top Ten Risk Behaviors in TV Shows, **A7**
Spot the Sales Pitch, **A30**
Analyze Relationships in Ads, **A47**

Make a Difference
Care Bags, **A20**

Careers
Psychologist, **A23**
Sociologist, **A71**

Health Online
Emotions and Health, **A25**
Resolving Conflicts, **A78**

UNIT B Growth and Nutrition

Learning Life Skills
Practice Healthful Behaviors, **B30**
Analyze What Influences Your Health, **B48**

Life Skills Activities
Manage Stress, **B9**
Be a Health Advocate, **B13**
Access Health Facts, Products, and Services **B17, B39**
Make Responsible Decisions, **B25, B60**
Analyze What Influences Your Health, **B29**
Use Resistance Skills, **B47**
Practice Healthful Behaviors, **B55**

Build Character
Make an Album, **B24**
Plan to Share Food, **B45**

Cross Curricular Links
Science, **B5, B54**
Physical Education, **B7**
Social Studies, **B23, B43**
Art, **B28, B38, B59**
Math, **B57**
Write About It!, **B12, B27, B58**

On Your Own for School or Home
Drink More Water, **B19**
Plan a Menu, **B44**

Consumer Wise
Analyze Ads, **B15**
Read the Labels, **B41**

Make a Difference
Ryan's Wells, **B53**

Careers
Nurse Practitioner, **B8**
Food Inspector, **B51**

Health Online
System Savvy, **B16**
Keeping It Safe, **B52**

UNIT C Personal Health and Safety

Learning Life Skills
Access Health Products, **C36**
Resolve Conflicts, **C72**

Life Skills Activities
Access Health Facts, **C5, C59**
Make Responsible Decisions, **C14, C65**
Practice Healthful Behaviors, **C21**
Set Health Goals, **C29, C47**
Use Communications Skills, **C35, C69**
Be a Health Advocate, **C51**
Use Resistance Skills, **C77**

Build Character
Perform a Skit, **C34**
Include Others, **C75**

Cross Curricular Links
Science, **C14, C43, C61**
Math, **C26**
Physical Education, **C32**
Social Studies, **C70**
Music, **C76**
Write About It!, **C17, C50, C63**

On Your Own for School or Home
See What Skin Does, **C6**
Avoid Eyestrain, **C12**
Make an Escape Plan, **C45**
Role-Play, **C55**

Consumer Wise
Make a Buying Guide, **C25**
Violence on TV, **C67**

Make a Difference
Safer Streets, **C49**

Careers
Personal Trainer, **C33**
Firefighter, **C44**

Health Online
Gear Up for Safety, **C31**
Get Weather Ready, **C56**

UNIT D Drugs and Disease Prevention

Learning Life Skills
Be a Health Advocate, **D22**
Manage Stress, **D70**

Life Skills Activities
Make Responsible Decisions, **D9, D59**
Set Health Goals, **D15, D65**
Analyze What Influences Your Health, **D21, D77**
Manage Stress, **D29**
Use Communication Skills, **D35**
Use Resistance Skills, **D39**
Practice Healthful Behaviors, **D49**
Access Health Services, **D55**
Be a Health Advocate, **D69**

Build Character
List Reasons to Respect Yourself, **D34**
Plan to Prevent Disease, **D47**

Cross Curricular Links
Science, **D13, D18, D61, D62, D76**
Art, **D17, D48**
Math, **D19, D67**
Music, **D20**
Write About It!, **D8, D12, D25, D33, D54, D74**

On Your Own for School or Home
Find Expiration Dates, **D7**
Fight Pressure, **D31**
Record Your Vaccines, **D53**

Consumer Wise
Design a Brochure, **D5**
Check for "Heart-Smart" Foods, **D64**
Write a Commercial, **D75**

Make a Difference
Gifts for Families, **D68**

Careers
Drug Counselor, **D27**
Public Health Worker, **D52**

Health Online
Explore Government Agencies, **D38**
Explore Chronic Diseases, **D73**

Features and Activities

UNIT E Community and Environmental Health

Learning Life Skills
Use Communication Skills, **E14**
Set Health Goals, **E38**

Life Skills Activities
Make Responsible Decisions, **E7, E43**
Access Health Facts, **E13**
Analyze What Influences Your Health, **E21**
Set Health Goals, **E27**
Use Communication Skills, **E35**
Practice Healthful Behaviors, **E49**
Be a Health Advocate, **E53**

Build Character
Clean Up a Park, **E20**
Volunteer to Help, **E52**

Cross Curricular Links
Science, **E8, E34**
Social Studies, **E17, E36**
Art, **E18, E48**
Math, **E47**
Write About It!, **E6, E23, E33, E43**

On Your Own for School or Home
Identify Recycled Products, **E19**
Design a Dream Job, **E25**
Diagram a Career, **E26**
Dispose of Hazardous Wastes, **E45**

Consumer Wise
Design a Brochure, **E11**
Design a Safety Poster, **E41**

Make a Difference
Reducing Trash, **E51**

Careers
Consumer Reporter, **E5**
Recycling Plant Worker, **E44**

Health Online
Budget Benefits, **E12**
Watch the Waste, **E42**

Life Skills

Life Skills are actions you can take to improve and maintain your health. The life skills that are taught in this text are listed below.

- Make Responsible Decisions, **A40**
- Use Resistance Skills, **A84**
- Practice Healthful Behaviors, **B30**
- Analyze What Influences Your Health, **B48**
- Access Health Facts, Products, and Services, **C36**
- Resolve Conflicts, **C72**
- Be a Health Advocate, **D22**
- Manage Stress, **D70**
- Use Communication Skills, **E14**
- Set Health Goals, **E38**

To help you learn these life skills, each of the Learning Life Skills features in this book includes Foldables™. Foldables™ are three-dimensional graphic organizers you will make. They will help you understand the main points of each life skill.

UNIT A

Mental, Emotional, Family, and Social Health

CHAPTER 1

Mental and Emotional Health, *A2*

CHAPTER 2

Family and Social Health, *A44*

CHAPTER 1

Mental and Emotional Health

Lesson 1 • What Are Health and Wellness? . . *A4*

Lesson 2 • Plan for a Healthy Life. *A10*

Lesson 3 • Your Personality and Character . . *A16*

Lesson 4 • Your Emotions *A22*

Lesson 5 • Taking Charge of Your Health . . . *A28*

Lesson 6 • Managing Stress *A34*

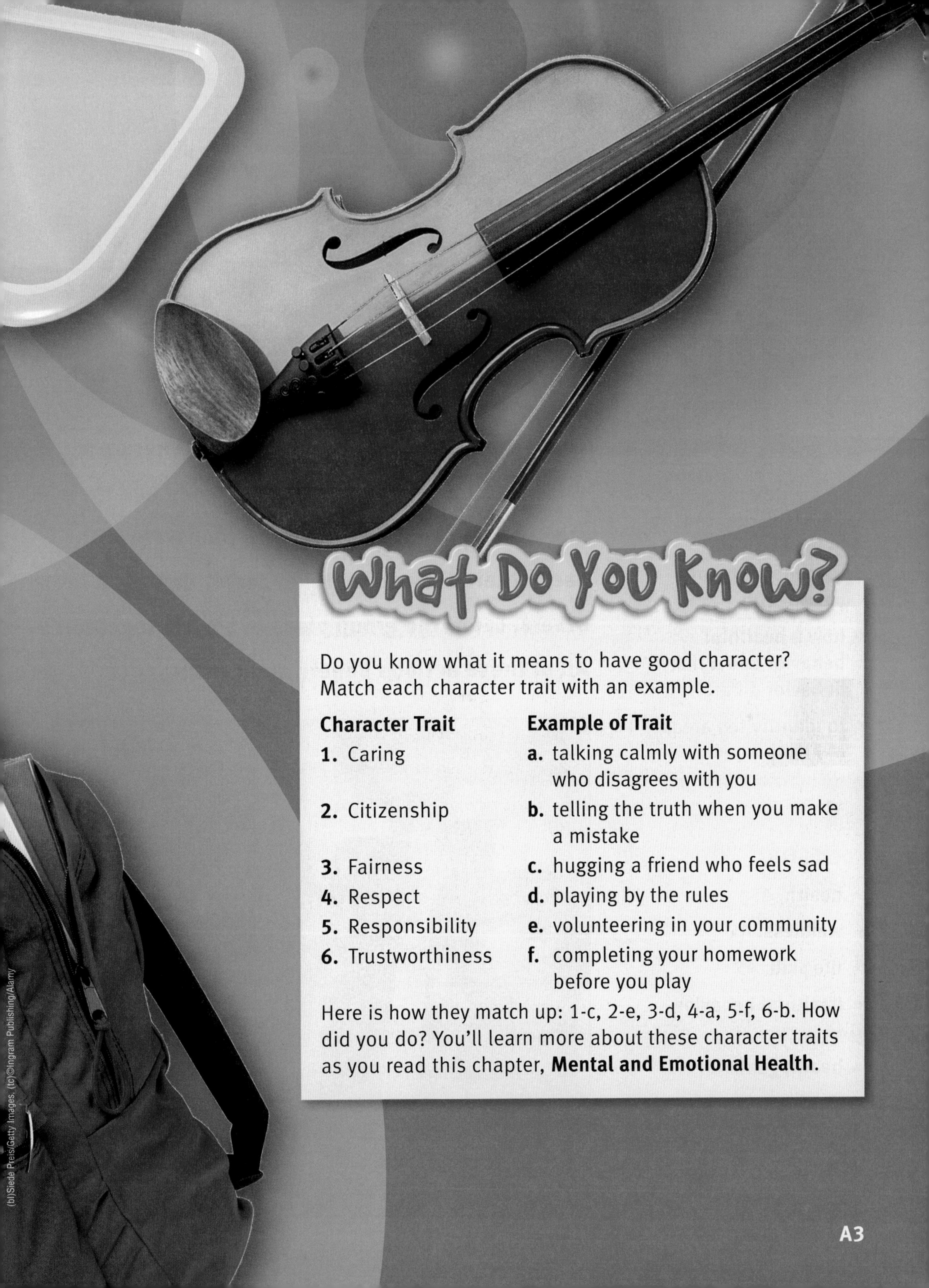

What Do You Know?

Do you know what it means to have good character? Match each character trait with an example.

Character Trait	Example of Trait
1. Caring	**a.** talking calmly with someone who disagrees with you
2. Citizenship	**b.** telling the truth when you make a mistake
3. Fairness	**c.** hugging a friend who feels sad
4. Respect	**d.** playing by the rules
5. Responsibility	**e.** volunteering in your community
6. Trustworthiness	**f.** completing your homework before you play

Here is how they match up: 1-c, 2-e, 3-d, 4-a, 5-f, 6-b. How did you do? You'll learn more about these character traits as you read this chapter, **Mental and Emotional Health**.

LESSON 1

What Are Health and Wellness?

You will learn . . .

- to identify ten life skills that can help you take responsibility for your health.
- how a healthful behavior and a risk behavior differ.
- to identify ten areas of health.

Vocabulary

- **health**, *A5*
- **wellness**, *A5*
- **life skill**, *A5*
- **healthful behavior**, *A6*
- **health goal**, *A6*
- **risk behavior**, *A7*

These young people are working together on a community project. As they work, they use their minds. They use social skills to cooperate with each other. They strive to do something good for others. Later, the group plans to bicycle together. All of these actions benefit their health.

Blend Images/Ariel Skelley/Getty Images

Health and Wellness

Health is the condition of your body, mind, and relationships. There are three parts to health.

Physical health is the condition of your body. It includes how well your body systems work. It also includes how you care for your body systems.

Mental and emotional health is the condition of your mind. It includes how you express your feelings and manage stress.

Family and social health is how you get along with others. It includes being a caring, responsible family member and friend. It also includes showing respect for others.

Wellness is the highest level of health you can achieve. You have wellness when you have health in all three areas mentioned above.

Life skills can help you achieve wellness. **Life skills** are actions that increase and maintain your health. You learn life skills now and practice them the rest of your life. The box at right lists ten important life skills you will read about.

Life Skills

- Practice Healthful Behaviors
- Set Health Goals
- Use Communication Skills
- Use Resistance Skills
- Make Responsible Decisions
- Analyze What Influences Your Health
- Access Health Facts, Products, and Services
- Be a Health Advocate
- Resolve Conflicts
- Manage Stress

How are health and wellness related?

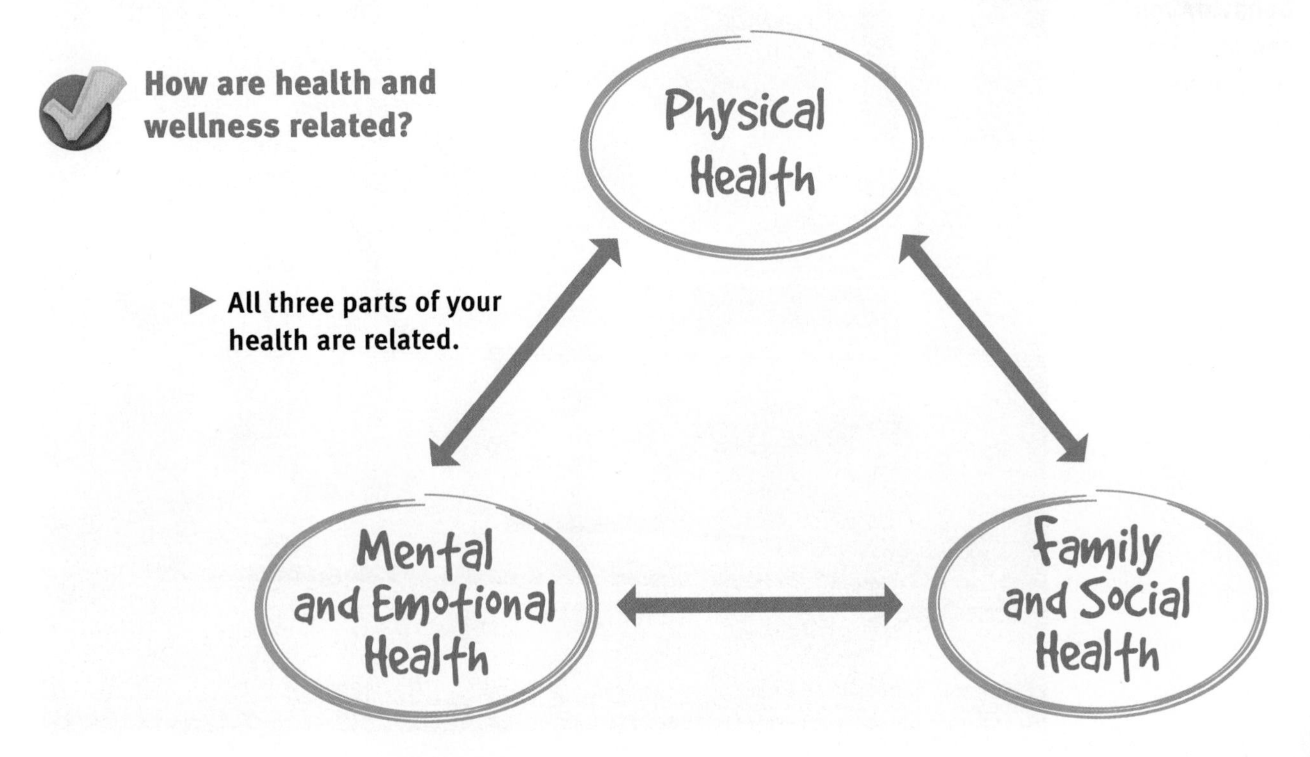

▶ All three parts of your health are related.

Practice Healthful Behaviors

Practice Healthful Behaviors

1. **Learn about a healthful behavior.**
2. **Practice the behavior.**
3. **Ask for help if you need it.**
4. **Make the behavior a habit.**

A **healthful behavior** is an action that increases the level of health for you and others. Practicing healthful behaviors is a key life skill.

Washing your hands before you eat is a healthful behavior. It removes germs from your hands. This reduces your risk of germs getting into your body and causing illness. Wearing a helmet when you ride a bicycle is a healthful behavior. A helmet protects your head from injury. Eating healthful foods is a healthful behavior. Healthful foods give your body energy. They help your body grow and develop.

A **health goal** is something that you work toward to achieve and maintain good health. Set a health goal to practice healthful behaviors.

CDC/Cade Martin

Washing your hands often is one healthful behavior you can practice. What others can you think of?

Avoid Risk Behaviors

Choosing healthful behaviors is one way to stay healthy. Avoiding risk behaviors and risk situations is another way. A **risk behavior** is an action that can be harmful to you and others. Riding double on a bicycle is a risk behavior. Another risk behavior is drinking alcohol. A *risk situation* is a situation that can harm your health. Someone who drives a car too fast creates a risk situation for the people in the car.

You can do something about risk situations. Choose healthful behaviors to keep some situations from becoming risk situations. For example, suppose a friend invites you to the beach. Choose healthful behaviors. Take steps to protect your skin from the sun's harmful rays. Apply sunscreen with an SPF of at least 15 and wear a wide-brimmed hat and sunglasses. Wear a long-sleeve shirt over your bathing suit to block the sun's rays.

List two examples of risk behaviors.

Consumer Wise ACTIVITY

Top Ten Risk Behaviors in TV Shows

Work with a group to analyze TV programs. With your parents' or guardian's permission, watch one TV show each night. List the risk behaviors you notice. Combine your list with your classmates' lists. Choose the ten behaviors you think are the riskiest for health. Rank them from one to ten, with #1 having the highest risk. Make a poster that shows your Top Ten List. On the back of your poster, make a Top Ten list of healthful behaviors. Present both sides of your poster to the class.

How could you make skateboarding a more healthful behavior? Technology is putting science to use. From studying the science of motion and of materials, technology has produced helmets and wrist guards that reduce the risk of injury.

Top Ten Areas of Health

Write About It!

Write Questions to Answer Choose one of the ten chapters of this book. Write three questions about the area of health covered in the chapter. They should be questions to which you don't know the answer. Keep your questions. When you finish the chapter you chose, see if you can answer them!

This book has information you need to help you practice life skills. It has ten chapters. Each one covers a different area of health.

- **Chapter 1, Mental and Emotional Health,** explains how to express feelings in healthful ways. It also lists ways to keep your mind sharp and manage stress.
- **Chapter 2, Family and Social Health,** discusses how to develop healthful relationships with friends and family members. It tells how to resolve conflict.
- **Chapter 3, Growth and Development,** describes how you change as you get older. It identifies healthful behaviors that help you grow and develop.
- **Chapter 4, Nutrition,** identifies food choices that promote health.
- **Chapter 5, Personal Health and Physical Activity,** discusses checkups and how to care for your body. It also identifies ways to keep your body fit.
- **Chapter 6, Violence and Injury Prevention,** describes safety rules to prevent injury. It teaches you first aid skills and ways to avoid violence and gangs.

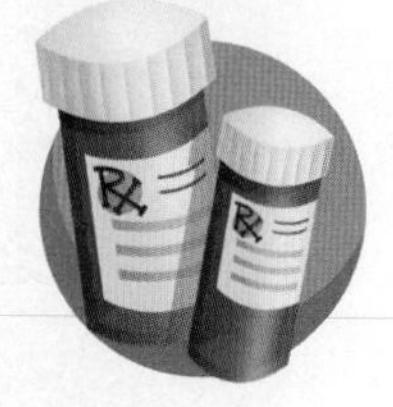

- **Chapter 7, Alcohol, Tobacco, and Other Drugs,** explains safe ways to use medicine. It teaches you about the harmful effects of some drugs.

- **Chapter 8, Communicable and Chronic Diseases,** describes the cause, symptoms, and treatments for diseases. It emphasizes how to prevent disease.

- **Chapter 9, Consumer and Community Health,** tells how to find health information. It gives you tips for choosing health products, too. You will learn about community health helpers and health careers.

- **Chapter 10, Environmental Health,** explains how you can help keep the air, land, and water clean and safe. It also describes how to conserve resources.

What information does the chapter on nutrition contain?

ACTIVITY

LIFE SKILLS

CRITICAL THINKING

Practice Healthful Behaviors

Choose a healthful behavior. You can choose a healthful behavior not listed in this lesson. Suppose you choose wearing safety gear.

1. **Learn about a healthful behavior.** Find the lesson in this book that includes the behavior you chose. Read about wearing safety gear. How does it keep you healthy?
2. **Practice the behavior.** For a week, make an effort to wear safety gear when it is necessary.
3. **Ask for help if you need it.** You might ask your parents or guardian to remind you to wear your safety gear.
4. **Make the behavior a habit.** Plan to do this behavior for three weeks. Keep track of how many times you did it. Is it now a habit?

LESSON REVIEW

Review Concepts

1. **Identify** five life skills that help you take responsibility for your health.
2. **Compare** a risk behavior and a risk situation. Give an example of each.
3. **List** three of the ten areas of health.

Critical Thinking

4. **Connect** Suppose you have a cold. This affects your physical health. Does it affect your mental, emotional, family, and social health, too? Explain why or why not.
5. LIFE SKILLS **Practice Healthful Behaviors** Make a list of three risk behaviors. Think of ways to change or replace each risk behavior with a healthful behavior. Explain how these changes can benefit your health.

LESSON 2

Plan for a Healthy Life

You will learn . . .

- why you need health facts and how to evaluate sources of health facts.
- how to set short-term and long-term health goals.
- how to make a health behavior contract.

Vocabulary

- short-term goal, *A12*
- long-term goal, *A12*

How can you improve your health? Start by setting a health goal. Then make a plan to work toward your goal. Using health knowledge can help you plan for a healthy life.

Health Facts

Suppose you want to start a program of regular physical activity. You will need health facts to decide how to do this. *Health facts* are true statements about health. To get health facts, you need reliable information. *Reliable* information is information that you can trust. You can get reliable information from people and groups who base their information on accurate scientific knowledge.

A coach, physical education teacher, trainer, and doctor would be qualified to give you health facts about physical activity. They have education and training in that area. They know about current research and scientific knowledge. They provide reliable information.

You might need health facts for another health topic. To make sure the information you get is reliable, ask the following questions.

- **What is the source of the health facts?** Suppose you read a billboard, magazine article, or something posted on the Internet. It is about diet and health. To decide if the information is reliable, first find out the source of the information. Always ask, "Who wrote this or said this?"
- **Is the source qualified to give this information?** Once you find the source, check out the source. Suppose the message on the billboard about diet and health came from the American Heart Association. This organization is a qualified source on the topic. A source that is qualified to give information on one area of health might not be qualified in another area.
- **Are the health facts based on current scientific knowledge?** The American Heart Association provides health facts based on current scientific knowledge. If you read something on the Internet, you might not be getting information that is based on up-to-date knowledge. Check out the source. Find out how they got their information.

Make Health Facts Yellow Pages

Choose a health topic you are interested in, such as bike safety. Identify three people or groups that can provide health facts on your topic. Make a page for each source listing the source's name and how to contact the source. Write a short paragraph telling why each person or group you chose is qualified to give reliable information.

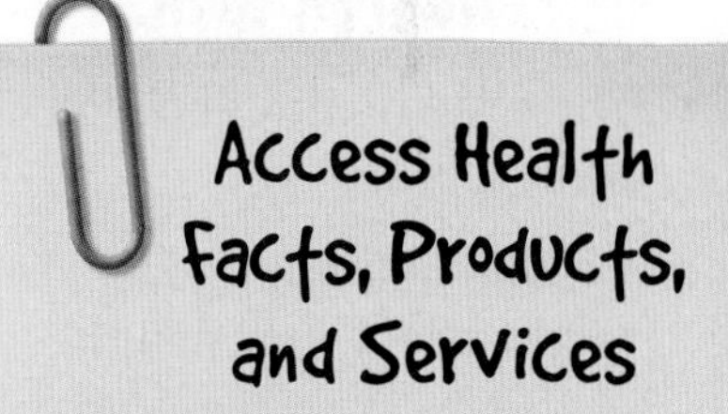

1. **Identify when you might need health facts, products, and services.**
2. **Identify where you might find health facts, products, and services.**
3. **Find the health facts, products, and services you need.**
4. **Evaluate the health facts, products, and services.**

Set Health Goals

ACTIVITY BUILD Character

Plan to Improve

Responsibility Talk to your parents or guardian about an area in which you could be more responsible. Then set a goal to do your best in that area. After a week, ask your parent or guardian how you have improved.

A health goal is something you work toward to achieve and maintain health. When setting a health goal, start with the words *I will*. This says that you take responsibility for reaching the health goal.

The amount of time it takes to reach a health goal varies. Health goals that can be reached in a short amount of time are called **short-term goals**. The health goal *I will get a medical checkup this year* could be a short-term health goal.

Long-term goals are goals that take a longer amount of time to reach. Some continue through your whole life. The goal *I will care for my body* is a long-term health goal. Short-term goals can be part of long-term goals. Getting a medical checkup this year is a step toward caring for your body systems.

It can be hard to work toward a long-term goal. Perseverance (puhr•suh•VIR•uhns) is important to reach a long-term goal. *Perseverance* means that you keep trying, even when it is difficult.

Setting a goal can help you stay focused on what you want to accomplish. Achieving your goal will give you a feeling of pride in yourself and boost your self-confidence.

Set Health Goals

1. **Write the health goal you want to set.**
2. **Explain how your goal might affect your health.**
3. **Describe a plan you will follow to reach your goal. Keep track of your progress.**
4. **Evaluate how your plan worked.**

▶ ***I will get plenty of physical activity* is a long-term health goal.**

Here are some health goals for each of the ten areas of health.

Health Goals

Mental and Emotional Health
I will take responsibility for my health.
I will show good character.
I will choose behaviors to have a healthy mind.
I will have a plan for stress.

Family and Social Health
I will show respect for all people.
I will work to have healthy family relationships.
I will settle conflict in healthful ways.
I will work to have healthy friendships.

Growth and Development
I will care for my body systems.
I will learn about the stages of the life cycle.
I will accept how my body changes as I grow.
I will be glad that I am unique.

Nutrition
I will eat healthful meals and snacks.
I will stay at a healthful weight.
I will read food labels.
I will use table manners.

Personal Health and Physical Activity
I will have regular checkups.
I will get plenty of physical activity.
I will get enough rest and sleep.

Violence and Injury Prevention
I will follow safety rules for my home and school.
I will protect myself from people who might harm me.

Alcohol, Tobacco, and Other Drugs
I will not drink alcohol.
I will say "no" if someone offers me a harmful drug.
I will not use tobacco.

Communicable and Chronic Diseases
I will choose habits that prevent the spread of germs.
I will recognize symptoms and get treatment for communicable diseases.

Consumer and Community Health
I will check out sources of health information.
I will choose safe and healthful products.

Environmental Health
I will help protect my environment.
I will not waste energy and resources.

Setting a health goal and working to achieve it shows that you are willing to take action when you need to. It shows that you have perseverance. Most importantly it shows that you value your health.

What are the two types of health goals?

Write an Explanatory Paragraph Choose one health goal from the box. Write a paragraph explaining how achieving the goal improves your health. Include steps you could use to achieve this health goal. What short-term goals could be part of the long-term goal?

Health Behavior Contract

A health behavior contract is a written plan for reaching a health goal. There are four steps to follow to make a health behavior contract.

Health Behavior Contract

Name ______________________ **Date** ________

1. **Write the health goal you want to set.**

1 Health Goal I will get plenty of physical activity.

2. **Explain how your goal might affect your health.**

2 Effect on My Health Physical activity helps make my heart and bones strong. It helps me keep a healthful weight. It helps reduce the harmful effects of stress.

3. **Describe a plan you will follow to reach your goal. Keep track of your progress.**

3 My Plan I will be physically active for at least thirty minutes each day. For one week I will record my activity on my calendar. I will put a star on each day of the week that I follow my plan.

My Calendar

Sun.	Mon.	Tues.	Wed.	Thurs.	Fri.	Sat.

Tell who can help ▲ you reach your goal—for example, a parent or guardian, a teacher, a school nurse, a doctor, or a coach.

4. **Evaluate how your plan worked.**

4 How My Plan Worked I will talk about how my plan worked. If I skip a day of activity, I will explain why.

ACTIVITY

LIFE SKILLS

CRITICAL THINKING

Set Health Goals

Read through the list of health goals on page A13. Choose one health goal that you would like to work toward. Then make a health behavior contract like the one on page A14.

1. **Write the health goal you want to set.** Write your health goal in the contract.
2. **Explain how your goal might affect your health.** Use your health knowledge. Discuss ways that your health goal will improve your health. Look through this book for ideas and ask your parent or guardian if you need to.
3. **Describe a plan you will follow to reach your goal. Keep track of your progress.** Write your plan into your Health Behavior Contract. Be sure you have a way to keep track of your progress. Share your Health Behavior Contract with a parent or guardian. Then follow it.
4. **Evaluate how your plan worked.** Decide when and how you will check to see if your plan works.

What are the four steps to follow when making a health behavior contract?

LESSON REVIEW

Review Concepts

1. **List** three questions to help you decide if health facts are reliable information.
2. **Describe** short-term and long-term health goals.
3. **Explain** how to make a health behavior contract for a health goal.

Critical Thinking

4. **Evaluate** Suppose you wanted to find out more about the disease chicken pox. What would be a reliable source of that information?
5. LIFE SKILLS **Set Health Goals** Suppose you are having trouble achieving a health goal. Who could help you decide how to change your plan?

LESSON 3

Your Personality and Character

You will learn . . .

- what aspects make up your personality.
- what six traits make up good character.
- what actions help you have a healthy self-concept.

Vocabulary

- **personality**, *A17*
- **respect**, *A17*
- **self-respect**, *A17*
- **responsible**, *A17*
- **good character**, *A18*
- **self-concept**, *A20*
- **self-esteem**, *A21*

You are a unique individual. You have your own thoughts and feelings. You have likes and dislikes. You make choices that show your personality and character. These individual aspects combine to make you who you are.

Your Personality

Your **personality** is a blend of your traits, talents, and actions. Your personality is what makes you unique. A *trait* is something specific about the way you look or act. For example, you might be quiet or you might talk a lot. You might be serious or easygoing. You might be tall or short. A *talent* is something you are good at. You might be good at dancing or playing piano. Sports such as soccer might be your talent.

Part of your personality involves your actions. Do you show respect for yourself and others? Are you responsible? To **respect** someone is to treat that person with dignity and consideration. **Self-respect** is thinking highly of yourself. You are **responsible** when you are able to be trusted to do what you say you will do. Being responsible also means you can depend on yourself to choose healthful behaviors. You follow rules, laws, and your family guidelines. You have a healthful personality when you make the best of your traits, develop your talents, and choose responsible actions.

Design a Shirt

Design a T-shirt with symbols that represent your personality traits and talents. Draw your design. Share it with a classmate. Explain what each symbol means.

What is personality?

When you help others, you are being responsible. This can help your self-respect.

Good Character

▲ Trustworthiness

An important part of your personality is your character. **Good character** is choosing actions that show respect for yourself and others. If you have good character, you demonstrate these six traits through your actions. When you work with others and with groups, show the traits of good character. Using these traits will help all of you reach common goals and get the job done.

Trustworthiness You are honest. You do not lie, steal, or cheat. Others know that they can rely on you to keep your promises.

Responsibility You are responsible. People can depend on you to do what you say you will do. You think before you act.

Respect Your words and actions show that you respect others. You treat others fairly. You are polite and considerate of other people's feelings.

▲ Responsibility

▼ Respect

Fairness You treat others as you expect to be treated. You share with others and take turns. You listen to others with an open mind.

Caring You show others that you care about them. You help those who are in need. You are kind. You forgive others when they make a mistake.

Citizenship You do your share to make your school, your community, and the world a better place. You are a good neighbor and help take care of the environment. You follow the rules of your family and your school. You follow laws in your community.

Good character shows in a person's actions. If you are responsible, you don't just *say* "I'm responsible." You *act* in responsible ways. You make responsible decisions, even in difficult situations.

Describe two actions that show others that you are responsible.

Write About It!

Write Newspaper Headlines Invent a name for a newspaper based on one of the six character traits. For example, your newspaper might be called *The Trustworthy Times* or *Responsible Reports*. Write a headline and interview a teacher who shows one of the character traits.

Caring

Fairness

Citizenship

Self-Concept

MAKE a Difference

Care Bags

Annie Wignall was 11 years old when she started the Care Bags Foundation. She heard about other children who had to leave their homes because they were being abused or harmed there. These children often had to leave in a hurry. They couldn't take anything with them. Annie started gathering Care Bags for these children. The bags hold books, games, and necessary supplies such as toothbrushes and soap. How does Annie's project show caring for others?

The way you think and feel about yourself most of the time make up your **self-concept**. Having a healthful self-concept allows you to work toward and achieve your goals in life. There are ways to develop a healthful self-concept.

- **Build on your strengths.** Are you good at math? You might ask your math teacher if you can do an extra-credit project. Spending time on your areas of strength helps you be the best that you can be.
- **Work to improve your weaknesses.** If you are not skilled at basketball, you might ask a talented friend for tips. You might set aside time to practice. Listen to criticism. Use it to find ways to build your skills.
- **Develop good character.** Choose actions that show you are trustworthy, respectful, caring, responsible, fair, and a good citizen. Showing good character allows others to respect you.

Write About It!

Write a story about a person who sets a high goal for himself or herself. Even though the goal is difficult to achieve, the person sticks with it until he or she reaches it. State the goal and describe the actions the person took to reach the goal.

▶ **Of what accomplishments are you most proud?**

SW Productions/Photodisc/Getty Images

Actions that give you a healthy self-concept benefit self-esteem. **Self-esteem** is a feeling of pride in yourself. When you are proud of yourself, you want to take better care of yourself. You respect yourself. You accept all parts of your personality. You set health goals. You work toward your goals, too. You avoid risk behaviors and practice healthful behaviors. You learn from your mistakes and overcome setbacks.

You can help encourage others to have a healthy self-concept. Praise your friends when they do well and work hard. Maybe you are good at writing but one of your friends struggles with it. Maybe that friend is better at math than you are. You can help your friend with writing. Your friend can help you with math. Together you can work on your strengths and weaknesses.

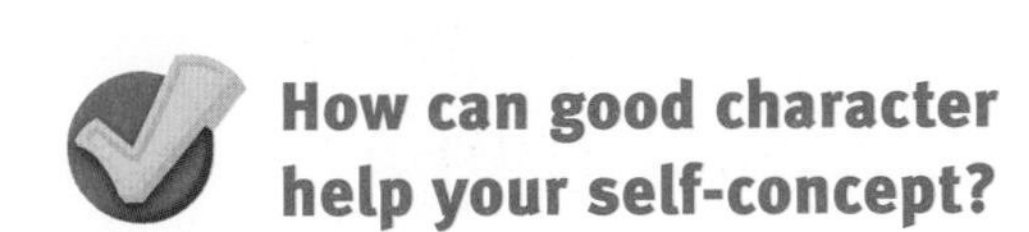

CRITICAL THINKING

Be a Health Advocate

You can encourage others to develop a healthful self-concept. Be an advocate for his or her good character.

1. **Choose a healthful action to communicate.** You know that having good character is important to your health.
2. **Collect information about the action.** Think about someone you know. Find out how he or she shows good character. Talk to your classmates.
3. **Decide how to communicate this information.** Make an award for the person you chose. Cut a star from paper. Label it "Good Character Award." On the back, write a few sentences explaining why the person deserves the award.
4. **Communicate your message to others.** Give the award to the person. Tell the person why he or she deserves it. Encourage your friends to do the same.

Good Character Award

LESSON REVIEW

Review Concepts

1. **Explain** three aspects of personality.
2. **List** the six traits of good character.
3. **Name** three actions you can take to keep a healthy self-concept.
4. **Identify** three actions that show a person has high self-esteem.

Critical Thinking

5. **Synthesize** Explain how a healthy self-concept helps you have a healthy personality.
6. LIFE SKILLS **Be a Health Advocate** A friend plays the piano. She is having a recital. What actions might you choose to advocate for a healthful self-concept for your friend?

LESSON 4

Your Emotions

You will learn . . .

- to identify different kinds of emotions.
- how to manage emotions such as depression and anger.
- what behaviors can help you have a healthy mind and prevent boredom.

Vocabulary

- **emotion**, *A23*
- **grief**, *A23*
- **shyness**, *A23*
- **anger**, *A23*
- **depression**, *A25*
- **boredom**, *A26*

How do you feel today? Are you happy or sad? Are you interested or bored? Emotions are a normal part of life. Some emotions make you feel good. Others can be hard to handle. Learning how to manage your emotions can improve all parts of your health.

Feelings and Emotions

Your feelings change all the time. You may be happy on the day of a friend's party. You may be excited as you plan to go to the party. You may be sad when you discover that your friend is ill and the party is canceled. An **emotion** is a feeling inside you. Understanding what you feel and why you feel that way is a big step toward mental and emotional health.

Emotions can be pleasant to feel. Scoring a soccer goal may make you feel joy. Emotions can also be difficult or uncomfortable to feel. The loss of a family pet may cause grief. **Grief** is discomfort that results from a loss such as death.

Shyness is an uncomfortable emotion. **Shyness** is a feeling of not being comfortable around other people. You might feel shy when you are in a room full of strangers. Some people experience shyness even when they are with people they know.

Suppose that your little brother breaks one of your toys. How does this make you feel? It probably makes you angry. **Anger** is the feeling of being irritated or annoyed. Anger can be a very strong emotion.

What emotions make you feel uncomfortable?

CAREERS

Psychologist

Psychologists (sigh•KAHL•uh•jists) are health care professionals who study how people think, feel, and behave. They use their knowledge to help people deal with uncomfortable situations. Psychologists work in hospitals, schools, and counseling centers.

▼ **A psychologist can help people learn to manage their emotions. What are other ways you can manage strong emotions?**

Managing Emotions

ACTIVITY

Art LINK

Rainbow of Feelings

Paint a picture to express anger, joy, love, or fear. The subject of the painting and the colors you use should express the emotion. Show your painting to the class. Can classmates guess what emotion is expressed?

Emotions are normal. It's okay to feel angry, shy, or sad. What's important is what you do about your emotions. Managing your emotions in healthful ways protects all areas of your health.

There are three questions you can use to help manage your emotions.

1. What am I feeling?
2. Why do I feel this way?
3. How can I express what I feel in a healthful way?

These questions can help you manage very strong emotions. If you feel angry, afraid, or shy, stop and think before you act. Use the questions to help you find a healthful way to express your emotions.

ACTIVITY

LIFE SKILLS CRITICAL THINKING

Use Communication Skills

Work with a classmate. Brainstorm situations in which you might feel fear or shyness. Then role-play the situations.

1. **Choose the best way to communicate.** Use the questions above to decide how to communicate your emotions.
2. **Send a clear message. Be polite.** Stay calm. Speak clearly.
3. **Listen to each other.** Be sure both of you communicate your emotions clearly and calmly.
4. **Make sure you understand each other.** If you aren't sure what emotion your classmate is communicating, ask him or her to try another way.

After you finish, switch roles and try again. Then work together to write a paragraph about the situation. Explain what worked and what didn't as you tried to communicate with each other.

Gary He/McGraw-Hill Education

Depression and Anger

Your mind and body are connected. Your thoughts and feelings affect your body. Managing your emotions helps your physical health.

Depression is a feeling of sadness and gloom. If you are depressed for a long time, the cells in your blood that fight disease may not work as well as usual. This could make you more likely to become ill. If you feel sad or depressed, use the questions on page A24. Try to understand why you feel the way you do. Talk to a responsible adult about your feelings.

Anger might be difficult for you to express. However, keeping anger inside can have physical consequences. When you stay angry, you might not sleep. The tension you feel might increase your heart rate and blood pressure. This increases your risk of disease.

Here are some strategies you can use to manage your anger in a healthful way.

- **Leave the situation** and cool off. Give yourself time to cool down before you talk with the person.
- **Count to ten** before you say or do anything. This gives you time to think about what you want to say or do.
- **Tell the person** why you feel angry. He or she might not be aware of how you feel.
- **Be physically active** to reduce stress and tension.
- **Ask an adult** for help. An adult might have a solution that you did not think of.

What you feel can affect your physical health. Research how your body reacts physically to strong emotions. Use the e-Journal writing tool to write a report on what you learn.

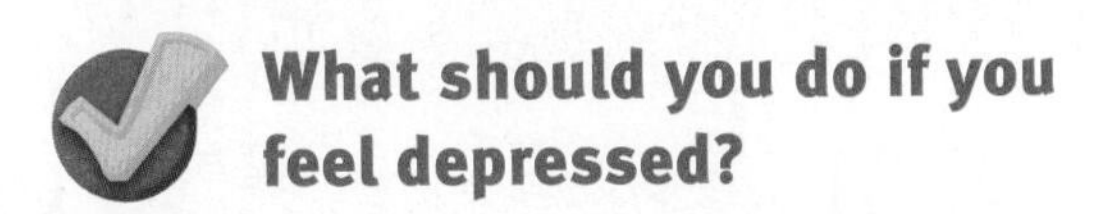

What should you do if you feel depressed?

▶ **Physical activity is a healthful way to relieve the physical effects of anger.**

Big Cheese Photo/SuperStock

A Healthy Mind

Write About It!

Fight Boredom Draw a pair of boxing gloves. Decorate with glitter, stickers, or pictures. Paste them on a shoebox and title the box *My Fight Boredom Box.* Cut 15 strips of paper. On each strip, write something that you can do when you feel bored. Exchange five strips with a classmate. Put the 15 strips in your box. What will you do the next time you are bored?

Another important part of mental and emotional health is having a healthy mind. You keep a healthy mind when you use it often. When you read, solve puzzles, play word games, and learn new things, you help keep your mind strong and healthy.

You also can keep your mind healthy by fighting boredom. **Boredom** is the state of being restless and not knowing what you want to do. Boredom makes you feel as if there is nothing to do.

You can fight boredom by trying new things. Join a school club or a sports group. Visit a local recreation center to find out about classes it offers. Ask a responsible adult for suggestions on how to develop a new interest. There are always new things to do. There's no reason to be bored!

▼ **Is there something you'd like to learn about? Learning new things is a fun way to keep a healthy mind.**

Keeping your body in top form helps keep your mind sharp, too. Physical activity helps blood move through your body. The blood carries oxygen to your brain. Your brain needs oxygen to work well.

Avoid alcohol, tobacco, and other harmful drugs. Alcohol and other harmful drugs make the mind less alert. Carbon monoxide in cigarette smoke prevents the brain from getting the amount of oxygen it needs to work well.

Make sure to eat a healthful breakfast. This gives you energy all morning long. Breakfast provides your brain with the energy it needs to work properly. You will be alert in school and ready to learn.

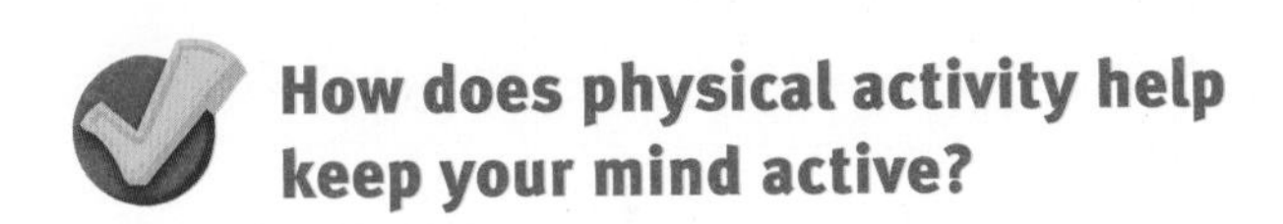

How does physical activity help keep your mind active?

Cut out pictures of breakfast foods from old newspapers, magazines, and food wrappers. Make a collage of breakfast foods by gluing the photos and wrappers to a paper plate. Share your collage with your parent or guardian. Plan a healthful breakfast with them.

LESSON REVIEW

Review Concepts

1. **Define** *emotion* and give three examples of strong emotions.
2. **Explain** how to manage anger.
3. **List** three things you can do to avoid boredom.

Critical Thinking

4. **Synthesize** How does expressing your emotions in healthful ways help your physical health?
5. LIFE SKILLS **Use Communication Skills** Suppose that a friend broke a promise to you. Write a note that you could send to your friend expressing your anger in a healthful way.

LESSON 5

Taking Charge of Your Health

You will learn . . .

- what steps you can take to make responsible decisions.
- what factors can influence your decisions, including peer pressure.
- how resistance skills can help you resist pressure to make a wrong decision.

Vocabulary

- **peer**, *A31*
- **peer pressure**, *A31*
- **resistance skills**, *A32*

You are in charge of your health. You choose actions that are either healthful or harmful. Many factors can influence your choices. Learning about those influences can help you make responsible health decisions.

Making Responsible Decisions

You make decisions all the time. Some decisions do not affect your health. Maybe you choose the clothes you wear to school or what color scarf to buy. Other decisions do affect your health. You might decide what to eat for lunch. You might choose what sport to play.

When you make a decision about health, it's important to make a responsible decision. A responsible decision is healthful and safe. It follows rules and laws. It shows respect for you and others. It follows family guidelines and shows good character.

There are steps you can follow to help you make responsible decisions. These steps are listed at the right. For example, suppose a friend invites you to ride bikes. You do not have a helmet with you.

First, identify your choices. You could choose not to ride bikes or choose to ride without wearing a helmet. Then evaluate your choices. Use the *Guidelines for Making Responsible Decisions*™. These guidelines include the following six questions.

- Is it healthful?
- Is it safe?
- Does it follow rules and laws?
- Does it show respect for myself and others?
- Does it follow family guidelines?
- Does it show good character?

If you answer "yes" to all of these questions, the choice is a responsible decision. Check your decision with your parent, guardian, or another trusted adult. Finally, evaluate your decision. How did it work out for you?

Write About It!

Review a Story Choose a story or novel that you like. Find a health decision that a character in the story made. Write a paragraph about the decision. Would you have made the same decision? Use your health knowledge to explain why or why not.

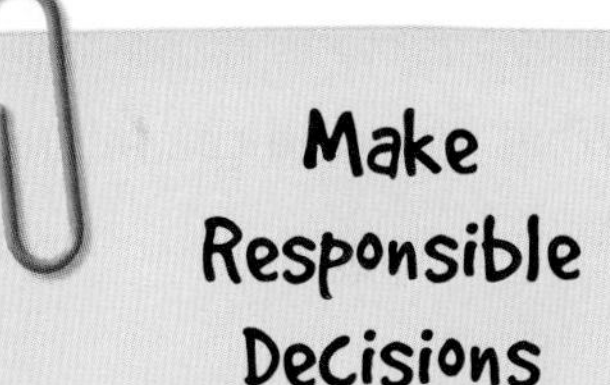

1. **Identify your choices. Check them out with your parent or trusted adult.**
2. **Evaluate your choices. Use the *Guidelines for Making Responsible Decisions*™.**
3. **Identify the responsible decision. Check this out with your parent or trusted adult.**
4. **Evaluate your decision.**

Influences on Your Health

Spot the Sales Pitch

Find three ads in newspapers or magazines. With permission, cut out the ads and glue them onto a large sheet of paper. Below each ad, write how the ad's maker is trying to influence you. Is there a famous person in the ad? Does the ad claim that the product will make you healthier? Present your results to your class.

Suppose you have to choose what to eat for lunch. What might influence your choice? An *influence* can lead you to act in a certain way. Knowing what influences your health can help you make responsible decisions. Some influences are healthful. Others may be harmful. When making health decisions, try to avoid harmful influences.

The following are all factors that could influence your choice of lunch.

- **Your personal preferences influence your behaviors and decisions.** Some influences come from your personality. You might prefer the taste of pizza to that of salad. Others come from health concerns. Perhaps you are allergic to nuts and can't eat peanut butter.
- **Other people might influence your behaviors and decisions.** Your family members strongly influence the choices you make. Perhaps your family usually eats soup at lunchtime. You may choose to eat soup also. Your friends also influence your choices. A friend might ask you to try a sandwich she likes.
- **Things you see and hear might influence your behaviors and decisions.** Ads that you see on television and hear on the radio can also affect your health decisions. Maybe a commercial shows your favorite athlete drinking a new juice. You might decide to try it because you like the athlete. Your friends might say that the athlete is a good role model.

▼ What you read can influence your health.

Tim Pannell/SuperStock

Peer Pressure

Peers, or people your age, include your friends and classmates. They may have a strong influence on the way you think, dress, and behave. The influence people your age have on you is called **peer pressure**.

Your peers might use *positive peer pressure* to influence you to make responsible decisions. A classmate may encourage you to study for a test, for example. Studying is a responsible decision.

Your peers might use *negative peer pressure* to influence you to make wrong decisions. Suppose a group of friends asks you to skip your chores and go to a movie. If you let your friends influence you to do this, you would not be making a responsible decision. Sometimes friends don't mean to apply negative peer pressure. They may not realize that you have chores to do. It is up to you to make this clear to them.

When your peers pressure you to do something wrong, you don't have to do it. You can choose not to go along with negative peer pressure. A healthy self-concept will help you have the confidence to resist negative peer pressure. Use the steps in the margin to analyze what influences your health.

Analyze What Influences Your Health

1. **Identify people and things that can influence your health.**
2. **Evaluate how these people and things can affect your health.**
3. **Choose healthful influences.**
4. **Protect yourself against harmful influences.**

What are three things or people that influence you?

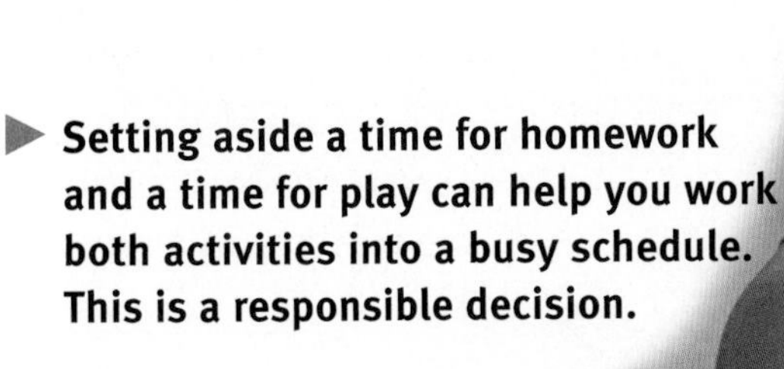

Setting aside a time for homework and a time for play can help you work both activities into a busy schedule. This is a responsible decision.

ACTIVITY BUILD Character

Citizenship Suppose a peer is pressuring a non-English speaking student to smoke cigarettes. With a partner, come up with ways the student could resist pressure to make a wrong decision. Emphasize different ways the student could communicate even though he or she speaks a different language.

Practice Resistance Skills

It's not always easy to say "no." Your friends may want you to do something harmful or unsafe. They may keep asking you even after you say "no." You may be tempted to make a decision that is not responsible.

The best thing to do is to avoid situations in which you know you would need to use resistance skills. If you know a person will pressure you to smoke cigarettes, for example, stay away from that person. You'll be less likely to feel pressure to act in ways that aren't responsible.

If this isn't possible, you can resist their pressure. **Resistance skills** help you resist pressure to make a wrong decision. Take the following steps when you use resistance skills.

- **Look at the person. Say "no" in a firm voice.** Be polite and calm, but don't back down.
- **Give reasons for saying "no."** Explain why you won't do what the person suggests.
- **Match your behavior to your words.** Use a calm tone and facial expression. Don't act as though you agree with the person if you don't. It's okay to walk away if the person will not stop pressuring you.
- **Ask an adult for help if you need it.** An adult can help you and give you support.

▲ You can use resistance skills to say "no" when peers pressure you to make a wrong decision.

Ken Karp/McGraw-Hill Education

ACTIVITY

LIFE SKILLS

CRITICAL THINKING

Use Resistance Skills

A good friend asks you to help play a mean trick on a classmate. You don't want to play the trick. How can you say "no"? Role-play the situation with a partner. Take turns playing each role. Use these steps when you are the person refusing.

1. **Look at the person. Say "no" in a firm voice.** Don't yell or shout. Speak clearly and calmly.
2. **Give reasons for saying "no."** Role-play different reasons you could give. You might say, "Playing that trick will hurt someone's feelings."
3. **Match your behavior to your words.** Role-play what you could do. You might choose to walk away from your friend.
4. **Ask an adult for help if you need it.** With your partner, decide what adults you might ask for help.

After you finish, make a list of reasons to say "no" to unhealthful behaviors. Put your list in a place where you will see it often.

LESSON REVIEW

Review Concepts

1. **List** the six questions you can ask yourself to decide whether a decision is responsible.
2. **Explain** how peer pressure can influence your health choices.
3. **Identify** the steps to follow when you need to use resistance skills.
4. **Describe** three factors that could influence your health choices. Give an example of each.

Critical Thinking

5. **Infer** Why might a peer pressure you to do something responsible? Why might a peer pressure you to do something that is not responsible?
6. **Summarize** Explain how both positive and negative peer pressure can influence you.
7. LIFE SKILLS **Use Resistance Skills** A classmate asks to copy your homework. You have tried saying "no." What should you do next?

LESSON 6

Managing Stress

You will learn . . .

- what stress is and how it affects your body.
- what steps to follow in managing stress.
- ways to bounce back from hard times.

Vocabulary

- **stress**, *A35*
- **stressor**, *A35*
- **eustress**, *A35*
- **distress**, *A35*

How busy is your life? Is your calendar packed with things to do? A busy schedule can cause a feeling that everyone has from time to time—stress! You can learn healthful ways to manage and cope with stress.

Everyday Stress

Stress is the response to any demand on your mind or body. Anything that causes stress is a **stressor**.

Suppose there's a test coming up at school. You may feel stress about the test. If you feel ready for the test, the stress may make you concentrate and do well. Stress caused by positive events is called **eustress** (YU•strehs).

If you're worried about the test, your body might respond in harmful ways. You might get a headache or a stomachache. You might feel tired and not do well on the test. These are examples of **distress**, which is stress caused by negative events.

Everyone experiences distress from time to time. But having too much distress or having distress too often can harm your health. Talk with a parent, guardian, or responsible adult if you experience distress often.

What is eustress?

Images-USA/Alamy

Effects of Stress
Heart rate and blood pressure increase
Lungs breathe in more oxygen
Blood vessels carry extra sugar for energy

Manage Stress

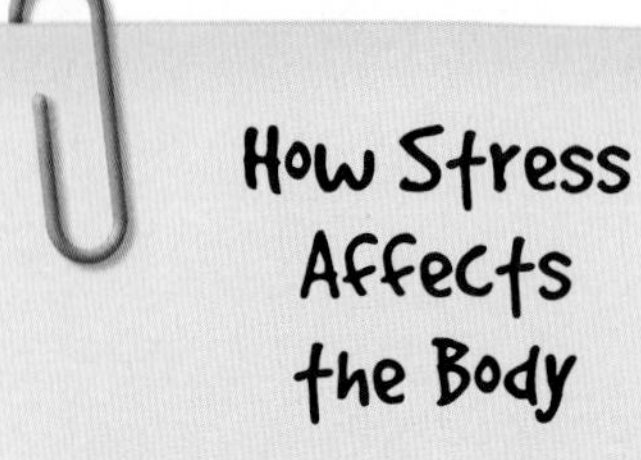

- ✔ **Extra sugar is released into the bloodstream for quick energy.**
- ✔ **Heart rate and blood pressure increase to get more blood flow to muscles.**
- ✔ **Pupils dilate to improve sight.**
- ✔ **Breathing rate increases to get more oxygen into the bloodstream.**

Stress management skills are ways to prevent stress and reduce its harmful effects on the mind and body. These steps can help you manage stress.

Suppose you have a championship soccer game this weekend. You have a stomachache and a headache. You have difficulty sleeping.

- **Identify the signs of stress.** You have a headache, a stomachache, and difficulty sleeping.
- **Identify the cause of stress.** The cause is your excitement and desire to perform well in the soccer game.
- **Do something about the cause of stress.** You could practice with a teammate. You could talk to your parents, guardian, or coach about the stress.
- **Take action to reduce the harmful effects of stress.** You might take a warm bath before bed. You might try relaxation exercises. You also might use some of the stress management skills listed on the next page.

If you feel stress for a very long time, talk to a parent or guardian. Your parent or guardian can get help from a doctor if you need it.

Daily Stress Busters

Here are some everyday activities that can help you reduce the stress you feel.

- **Plan your time wisely.** Try not to pack too many activities into one day. This helps you from feeling overwhelmed.
- **Get plenty of physical activity.** Physical activity can reduce the tension in your muscles and help you relax. It uses up the extra sugar that is released into the bloodstream when you have stress.
- **Eat healthful foods.** Food provides energy. When you feel stress, your body needs good fuel more than ever. Stress uses up vitamins B and C. Eat foods such as lean meats, nuts, enriched bread, strawberries, greens, and oranges to get more of these nutrients.
- **Get plenty of sleep.** Sleep helps keep the body from being tired. It is hard for a tired body to fight germs that cause disease. Sleep also helps your heart rate and blood pressure slow down.
- **Avoid foods and drinks that contain caffeine.** Soda, coffee, tea and chocolate are all high in caffeine.
- **Talk to your parents, guardian, or another responsible adult.** Talking about stress reduces its harmful effects. Others can respond with support and encouragement.
- **Use breathing exercises.** This simple breathing exercise slows the heart rate, blood pressure, and breathing rate. With your mouth closed, breathe in gently through your nose for a count of 4. Hold your breath for a count of 4. Then slowly breathe out through your mouth for a count of 4. Repeat this exercise three to four times.

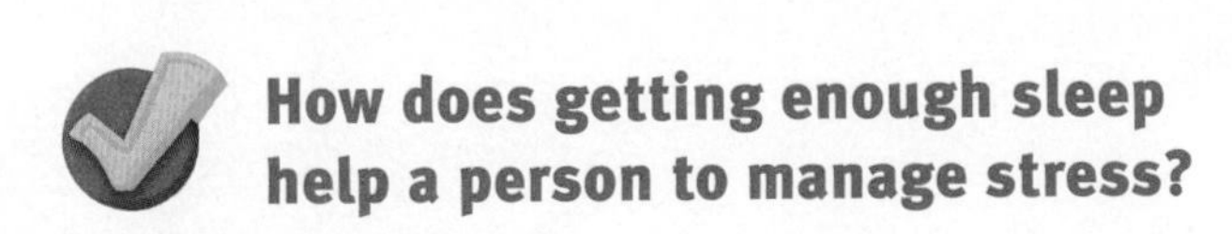

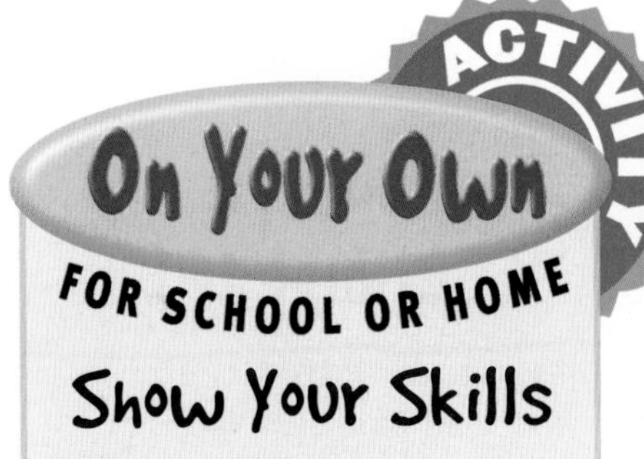

Show Your Skills

With your parents or guardian, discuss different ways to manage stress. Brainstorm actions that both of you could take to manage stress. Make a list of the actions. Display the list in a place where you will see it often.

Coping Strategies

Write About It!

Write an Expressive Poem
Write a poem about someone bouncing back from a hard time. In your poem, write about how the person felt. Describe how he or she dealt with the situation. Use words that express emotions.

Things do not always go the way you want them to. How would you feel if you tried out for the school play and didn't get chosen? What if your friend didn't invite you to her birthday party?

It's normal to feel disappointed when things like this happen. You don't have to pretend that you're happy when you're disappointed or sad. What's important is how you react to those feelings. How can you bounce back from hard times?

Your *attitude,* or way of thinking and seeing things, affects response to hard times. If your attitude is positive, you look on the bright side of things. You accept failure and move on. You might look for another play to try out for. Maybe you can find a community drama club to join. There, you might meet some new friends! With a positive attitude, you focus on what you do have and can do.

▼ Creative activities can help a person express feelings and bounce back from hard times.

Even when you keep a positive attitude, there will be times when you feel down. It's important for you to develop a support network for those down times. Your *support network* is a group of people who care about you. Family members and good friends make up your support network. These are people you can turn to during tough times. They are people with whom you can share your thoughts and feelings.

Let your support network help you. There will be a time when you will be able to return the favor as part of another person's support network. If you have a friend who needs your support, show the friend that you care. Listen if the friend needs to talk.

A person should never turn to risk behaviors such as smoking or drinking alcohol to deal with hard times. They'll make the hard times worse, not better. Ask your support network for help instead.

ACTIVITY LIFE SKILLS **CRITICAL THINKING**

Manage Stress

With a group, make a poster to show others how to manage stress. Put these steps on your poster.

1. **Identify the signs of stress.** List body changes that are signs of stress.
2. **Identify the cause of stress.** Include pictures of events that can cause stress. You might show someone taking a test or trying out for a band.
3. **Do something about the cause of stress.** Describe healthful ways to deal with stressful situations.
4. **Take action to reduce the harmful effects of stress.** Show ways that you can protect your body and mind from stress.

Hang your poster in your classroom or school library.

LESSON REVIEW

Facts and Skills

1. **Identify** what happens in the body when you feel stress.
2. **List** the steps to follow to manage stress.
3. **Describe** how you could help a friend bounce back after hard times.

Critical Thinking

4. **Explain** How do breathing exercises reduce the harmful effects of stress?
5. LIFE SKILLS **Manage Stress** Your brother is stressed about an upcoming test. Write a letter to him giving him advice on how to manage his stress.

Learning LIFE SKILLS

CRITICAL THINKING

Make Responsible Decisions

Problem You're about to ride your bike home from a long soccer practice. A teammate asks to ride double with you. You want to help your teammate. You know riding double can be dangerous, though. What should you do?

Solution Use the steps on the next page to help you make a responsible decision.

Learn This Life Skill

Follow these steps to help you make a responsible decision.

1 Identify your choices. Check them out with your parent or trusted adult.

What could you do? You could let your friend ride double. You could say "no." What other choices do you have?

2 Evaluate each choice. Use the *Guidelines for Making Responsible Decisions™*.

For each choice, ask yourself the following questions:

- Is it healthful?
- Is it safe?
- Does it follow rules and laws?
- Does it show respect for myself and others?
- Does it follow family guidelines?
- Does it show good character?

3 Identify the responsible decision. Check this out with your parent or trusted adult.

If you answer "yes" to all six questions, the decision is responsible. What choice would be responsible?

4 Evaluate your decision.

When you make a decision, you have to live with it. What might happen if you say "no"? What might happen if you let your friend ride double?

Practice This Life Skill

With a small group of classmates, think of another situation in which you might have to make a tough decision. Role-play how you could use the steps to make a responsible decision.

CHAPTER 1 REVIEW

Use Vocabulary

anger, *A23*
good character, *A18*
emotion, *A23*
health goal, *A6*
peer pressure, *A31*
responsible, *A17*
stressor, *A35*
wellness, *A5*

Choose the correct term from the list to complete each sentence.

1. The feeling of being irritated or annoyed is __?__.
2. The highest state of health you can reach is __?__.
3. Something that you work toward to become a healthier person is a(n) __?__.
4. A strong feeling such as sadness or happiness is a(n) __?__.
5. Choosing actions that show respect for yourself and others shows __?__.
6. Anything that causes stress is a(n) __?__.
7. When other people can depend on you to do what you say you will, you are __?__.
8. The influence people your age have on you is called __?__.

Review Concepts

Answer each question in complete sentences.

9. What is a short-term goal? A long-term goal?
10. What is the difference between positive and negative peer pressure?
11. List six traits that show good character. Tell one thing you can do to show each.
12. Name four steps to follow to manage stress.
13. Define a risk behavior. Give an example.
14. What are three ways to develop a healthful self-concept?

Reading Comprehension

Answer each question in complete sentences.

Emotions are normal. It's okay to feel angry, shy, or sad. What's important is what you do about your emotions. Managing your emotions in healthful ways protects all areas of your health. To help yourself manage strong emotions, you can ask yourself the following three questions. What am I feeling? Why do I feel this way? How can I express what I feel in a healthful way?

15. What are some emotions you might feel?
16. How can you protect your health?
17. How might you manage strong emotions?

Critical Thinking/Problem Solving

Answer each question in complete sentences.

Analyze Concepts

18. How does a Health Behavior Contract help you reach health goals?

19. Your friend encourages you to go in-line skating even though you forgot your helmet. What should you do?

20. How does your personality make you unique?

21. An advertisement says that a new sports drink is the best one for kids. Is this a reliable source of information? Why or why not?

22. How can showing respect for others help you be a better friend?

23. Why is it important to get information from reliable health sources?

Practice Life Skills

24. **Make Responsible Decisions** You have a library book that is due this afternoon. A friend asks you to play soccer after school. If you do, you won't be able to get to the library before it closes. You'll have to wait until tomorrow to return the book. What should you do? Use the *Guidelines for Making Responsible Decisions*™ to help you decide.

Read Graphics

Look at the Venn diagram. It shows what Bailey did after school this week. Use it to answer the following questions.

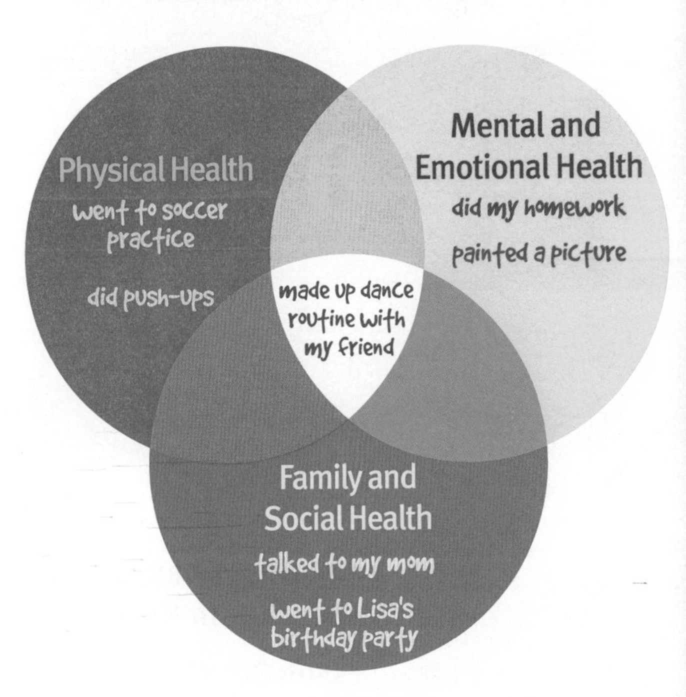

25. Did Bailey reach a balance between activities that helped her physical health, her mental and emotional health, and her family and social health?

26. Could any of the activities go in the overlapping sections? Which ones?

27. Why is "made up a dance routine with my friend" in the center of the Venn diagram?

CHAPTER 2

Family and Social Health

Lesson 1 • Your Social Health *A46*

Lesson 2 • Communication in Relationships. *A52*

Lesson 3 • When Conflict Occurs *A58*

Lesson 4 • Health in the Family *A64*

Lesson 5 • Facing Family Challenges *A68*

Lesson 6 • Among Friends. *A74*

Lesson 7 • Facing Challenges in Relationships. *A80*

What Do You Know?

This chapter focuses on relationships. Healthful relationships can improve your life in many ways. Check out what you know about healthful relationships. Write **true** or **false** for each statement.

__?__ There are many kinds of healthful relationships.

__?__ People in healthful relationships respect each other.

__?__ People in healthful relationships accept each other's differences.

__?__ Healthful relationships are free of conflict.

__?__ Helping each other stay healthy is part of a healthful relationship.

The fourth statement is **false**. Are you surprised that conflict can be part of a healthful relationship? In **Family and Social Health**, you will learn how to resolve conflict in relationships. You will also learn what makes a relationship healthful and how to make your relationships stronger.

LESSON 1

Your Social Health

You will learn . . .

- about different kinds of relationships.
- how to show respect and earn the respect of others.
- ways to be an advocate for health.

Vocabulary

- **relationship**, *A47*
- **mutual respect**, *A49*
- **health advocate**, *A50*
- **role model**, *A50*

Like most people, you probably spend most of your day with other people. If you get along well with others, you help your health. You help other people's health, too.

Kinds of Relationships

Your social health is based on relationships. A **relationship** is a connection between you and another person. There are many kinds of relationships. Some are close and loving. Others are more limited. These relationships are all important to your health.

Family A healthy family shares loving relationships. Your family probably gives you love, support, and meets needs such as food, shelter, and clothing. A healthy family supports all its members. Your family helps you develop in healthful ways. Your parents or guardian teach you how to behave in ways that show good character.

Friends Your friends help you learn who you are apart from your family. You learn by exploring mutual interests. You support each other. You may also learn from friends of different cultures, with backgrounds from other parts of the world. Friends can influence each other in positive ways.

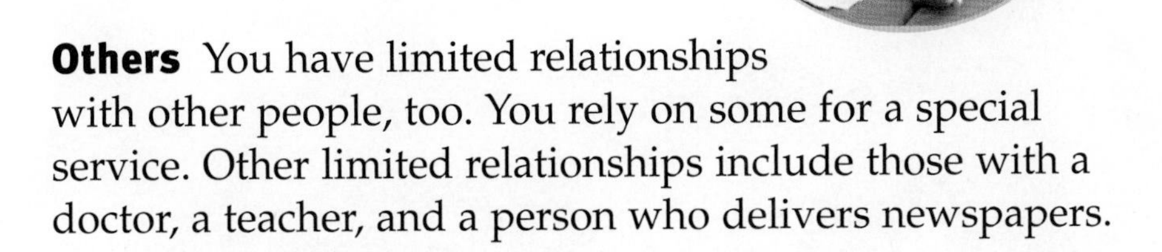

Others You have limited relationships with other people, too. You rely on some for a special service. Other limited relationships include those with a doctor, a teacher, and a person who delivers newspapers.

What kind of relationship would you have with a librarian or a dentist?

Analyze Relationships in Ads

The next time you watch television or read a magazine, check how the ads show relationships. Companies show relationships in their ads. The relationships and behaviors they show may not be healthful. Make a list of the relationships you see. Write whether you think each relationship is healthful and why.

Begin with Respect

Write About It!

Write a Persuasive Letter Suppose a student has been making fun of people who are different from him or her. Write a letter to persuade the student to respect others' differences.

People can be different from each other in many ways. They may speak different languages. They may have different hair or skin colors. They may have different beliefs and interests. Some people have special needs, such as a wheelchair or hearing aid. But all people want to be treated with respect. Treating people with respect shows that you value them for who they are.

You may meet people who are different from you. Show these people respect and you may be able to form a strong relationship.

Here are ways you can make them feel welcome.

- **Start a conversation** with them.
- **Invite them into a group** to which you belong, such as a club or sports team.
- **Introduce them** to your friends.
- **Ask them to join you** in an after-school activity.

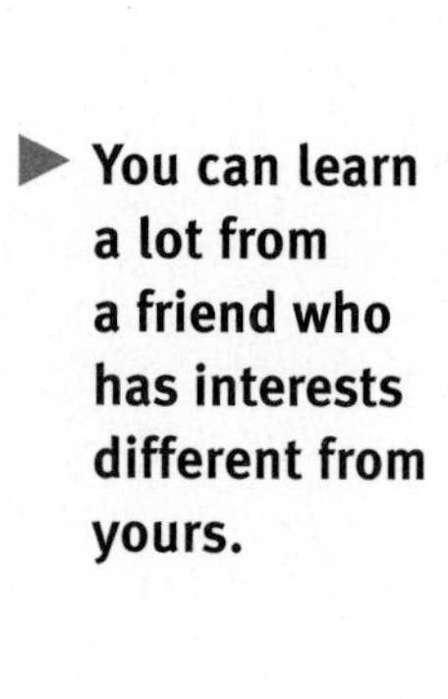

▶ **You can learn a lot from a friend who has interests different from yours.**

Fancy Collection/SuperStock

Expect Respect

Mutual (MYOO•chu•wul) **respect** is the high regard two people have for each other. It means that you show respect for others. It also means that others show you respect in return. Mutual respect is the foundation of any strong relationship.

To earn the respect of others, treat others with respect. When there is mutual respect, you and the other person see each other as equally important. Neither of you pushes the other around or puts the other down. You both talk and you both listen. You are open and honest in sharing your feelings.

▼ **Show respect for others. Expect respect from others. These are two basic tips for work with individuals and with groups. Respect for each other helps you work together to reach common goals.**

What is mutual respect?

ACTIVITY

Physical Education LINK

Try a Line Dance

When people dance together, they need to stay in step with each other. This shows respect because it prevents people from bumping into each other. Try this line dance to see how you can show respect.

1. Stand in a line with your classmates.
2. Step right with your right foot. Bring your left foot to your right foot. Repeat this step.
3. Step left with your left foot. Bring your right foot to your left foot. Repeat this step.
4. Step back with your right foot, then with your left foot. Step back with your right foot again. Then touch your left foot to the floor next to your right foot.
5. Step forward with your left foot, then back with your right foot. Then touch your left foot to the floor in front of your right foot.
6. Step forward on your left foot. Turn on your left foot one-quarter of the way around. Swing your right foot to make it line up with your left foot to complete the turn.
7. Repeat.

After you dance, talk with your classmates. How did it show respect to stay in step with each other? What would happen if someone did not stay in step?

Be a Health Advocate

ACTIVITY

Art LINK

Make a Thank-You Card

Make a thank-you card for a person who is a role model for you. Draw a picture of the person on the front of the card. Inside, explain how the person has influenced you in healthful ways. If you can, give your card to your role model. Thank him or her for being a health advocate.

A **health advocate** (AD•vuh•kut) is a person who helps keep others healthy. Some people may become health advocates because someone they care about chooses a risk behavior. Others may want to help people in their communities.

As a health advocate, you can encourage people to take responsibility for their health. Suppose you have a friend who eats snacks with lots of fat and sugar and who doesn't exercise. You might offer your friend healthful snacks. You might invite your friend to exercise with you. You might act as a role model.

A **role model** is a person whose behavior other people copy. You can be a role model. When you practice healthful behaviors, other people may decide to follow your lead.

▶ **Practicing a healthful behavior is the best way to encourage others.**

How to Be a Health Advocate

1. **Choose a healthful action to communicate.** Think about the people you know. Which healthful actions would help them?
2. **Collect information about the action.** Learn about the healthful action you choose. Go to the library or talk to people who know about the action. Decide which information will be most helpful for the people you know.
3. **Decide how to communicate this information.** One way to do this is to practice the healthful action yourself. You can be a good role model. You can tell people what you have learned. You can write letters. You can make posters to share with others. There are many ways to be a health advocate.
4. **Communicate your message to others.** Give other people the information they need to practice healthful actions.

How can you communicate health information?

ACTIVITY LIFE SKILLS CRITICAL THINKING

Be a Health Advocate

Work with a group to tell others about a life skill. Use the steps listed at the left.

1. Select a life skill from Chapter 1. Pick one that you want to tell others about.
2. Form a group with other classmates who selected the same life skill. Review the steps for the life skill. Why is this skill important? How will it help others to have a more healthful life?
3. List ways that you could tell others about the life skill. How will you communicate the steps?
4. Follow your plan. Make a four-panel cartoon illustrating the steps of the skill in action. Share your cartoon with other groups.

LESSON REVIEW

Review Concepts

1. **List** the three main kinds of relationships.
2. **Describe** how to show respect for others and earn their respect.
3. **List** the steps you can use to be a health advocate.

Critical Thinking

4. **Infer** Why might a person want to help improve the health of people in his or her community?
5. LIFE SKILLS **Be a Health Advocate** How might you communicate the benefits of physical activity to a friend?

LESSON 2

Communication in Relationships

You will learn . . .

- how to communicate in healthful ways.
- ways to use nonverbal communication and listening skills.
- ways to communicate emotions.

Vocabulary

- **communicate**, *A53*
- **I-messages**, *A53*
- **body language**, *A54*

Communicating well can help you build strong relationships with family and friends. It isn't easy to learn to communicate well, though. You have to know what you want to say. You also have to know how to say it in a clear, kind way.

Healthful Communication

To **communicate** means to exchange or share ideas, opinions, information, and feelings. Healthful communication can help you develop strong and trusting relationships. But effective communication is not always easy. You and another person may not agree on the best decision to make. Emotions can also make communication difficult. You might not want to talk about why you are sad. You might get so angry that you want to scream and say something mean.

It's important to make time to talk. It's also important to discuss each other's point of view. You can use I-messages to express how you feel in a respectful way. An **I-message** is an effective way to communicate feelings by stating the problem and how it affects you. An I-message names a behavior or situation. It also tells how the behavior or situation affects you. Using I-messages helps you avoid putting someone else down or placing blame. Here are two examples.

- I feel angry when you go into my room without asking me first.
- I feel that you don't care about me when you don't help me clean up.

The *Checklist for Communication* at the right lists some other tips to help you communicate well.

How can you avoid putting someone down when you are angry?

▶ **Listening is as important as talking.**

Checklist for Communication

These behaviors can help you communicate in clear and respectful ways. They are basic behaviors for working with individuals and with groups.

- **Speaking loudly enough** for other people to hear you
- **Speaking clearly** and choosing your words carefully so people understand you
- **Expressing feelings** in healthful ways
- **Acting in ways that match** what you say, so you do not confuse others
- **Listening carefully** when others speak, so you understand them
- **Watching the actions** of people speaking to you, because people communicate through actions as well as words
- **Making sure others know** that you are listening to what they have to say
- **Remembering what is said** to you
- **Repeating what others say** to make sure you understood them
- **Not telling secrets** to other people (unless the secret is about something that is not safe)

Nonverbal Communication

Write About It!

Write a Book Review Think about the characters in a book you read recently. How did they communicate their thoughts and feelings? Write a review of the book. Evaluate how people in the book communicated.

Communication involves more than speaking. Communicating without words is called *nonverbal* communication. **Body language** is the movements or gestures people make when communicating with another person. You may say more through your body language than through your words. For example, you may look worried, but say "I'm fine." This type of mixed message can confuse the listener.

Facial expressions reveal a lot about how you feel. What would you think if a friend raised his eyebrows when you asked a question? What might a frown or a blank look mean?

Posture is the way you hold your body. What posture shows that you feel confident? What posture shows that you are sad? You gesture using your head, arms, and hands. You may nod in agreement. You may clench a fist in anger even if you don't plan to use it.

▲ **Which of these people feels happy? Confused? Worried?**

Active Listening

When you communicate, listening is as important as talking. You can improve your listening ability with practice.

Active listening includes *attending*, or paying attention. You focus on what the speaker says. You look directly at the speaker. Active listening also includes acknowledging what you hear. You can *acknowledge* the speaker by repeating what you have heard in your own words.

You can use active listening with someone whose opinions may differ from yours. You may use these skills when talking with a friend who speaks little English. These skills can help you talk with someone who has a speech or hearing problem, too.

What are the two main parts of active listening?

Tips for Active Listening

- **Make eye contact.** This shows that you are paying attention to what the person is saying.
- **Use gestures.** Nod to show that you are interested.
- **Lean toward the speaker.** Let him or her know that you are paying attention.
- **Don't interrupt.** Wait to respond until the other person finishes speaking.
- **Repeat what was said to be sure you understand it.** Ask questions if you do not understand.

Communicating Emotions

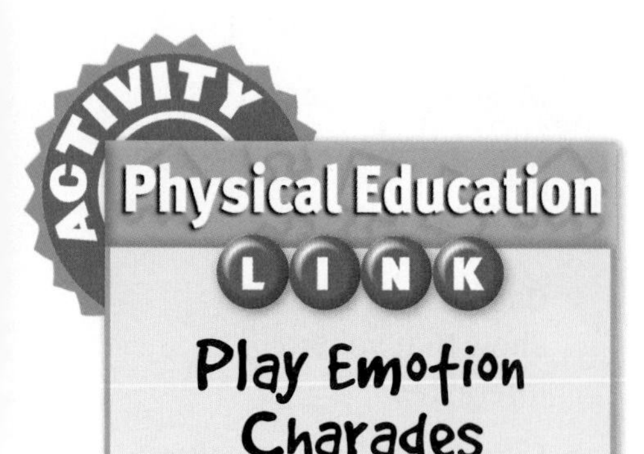

Play Emotion Charades

Write the names of emotions on pieces of paper. Turn the papers face down. On your turn, choose one. Try to communicate the emotion on the paper without talking. Use body language. Can your classmates identify the emotion?

Do you know what you are feeling right now? Does your body language show your emotions? Naming your feelings can help you communicate them.

You should feel comfortable expressing most emotions. You might hug a friend to show your gladness. You might share sadness with a friend over the loss of a pet. Sharing your emotions in these ways strengthens your relationships. At times, you might not want to share or show your emotions. That's normal. You don't have to share all your emotions. But sometimes people can't express their emotions even though they want to. This can cause stress. As you learned in Chapter 1, stress can harm your health.

Some emotions are strong. You may need to learn to manage these emotions. Managing emotions means expressing them in a healthful way. This can strengthen your relationships.

▲ **What emotions are expressed on these faces?**

Managing Strong Emotions

Self-control is the key to managing strong emotions. Using self-control can help you avoid saying or doing something you might regret later. You can communicate your emotions more clearly when you are calm.

Many people have trouble expressing anger. Try these tips if you feel angry.

- **Stop.** Take a break from what you are saying or doing.
- **Take time out.** Cool down.
- **Think about the situation.** Choose a healthful way to respond.
- **Act on your decision.** Use I-messages to talk about the situation.

What is the key to managing strong emotions?

ACTIVITY — LIFE SKILLS — CRITICAL THINKING

Use Communication Skills

Work with a small group. Each group member should think of a situation in which you would need to communicate. For example, you may need to talk to a teacher about a missed assignment, or to comfort a sibling who has broken a toy. Write your idea on a slip of paper. Fold the slips and put them in a pile. Take turns choosing a slip and acting out the situation.

1. **Choose the best way to communicate.** When it's your turn, decide how to use verbal and nonverbal communication.
2. **Send a clear message. Be polite.** Show your group how you would be very clear.
3. **Listen to each other.** Make sure each group member pays attention to the others.
4. **Make sure you understand each other.** Tell what emotions you think each group member communicates. Discuss how each of you could communicate more clearly.

LESSON REVIEW

Review Concepts

1. **List** three ways a person can communicate healthfully.
2. **Describe** how to use active listening skills.
3. **Explain** how to manage strong emotions such as anger.

Critical Thinking

4. **Synthesize** How might holding your emotions inside harm your physical health?
5. LIFE SKILLS **Use Communication Skills** Suppose you get a puppy. How could you communicate your joy without using words?

LESSON 3

When Conflict Occurs

You will learn . . .

- why and how conflict develops.
- what steps to take to resolve conflict.
- how a responsible adult can help mediate a conflict.

Vocabulary

- **conflict**, *A59*
- **prejudice**, *A59*
- **stereotype**, *A59*
- **peace**, *A60*
- **violence**, *A60*
- **mediation**, *A62*

No two people agree on everything. Conflict will creep into your relationships with your family and friends. You can find ways to resolve conflict peacefully.

How Conflict Develops

Conflict is a strong disagreement or fight. *Inner conflict* is a conflict within a person. You've probably felt inner conflict when you had to make a tough choice. Feeling jealous or hurt can cause inner conflict.

Conflict between people is called *interpersonal conflict.* Maybe you disagreed with a friend about what movie to see. Maybe you and your sister argue about who has to take out the trash.

Conflict can also occur because of cultural differences and prejudice. **Prejudice** (PRED•ju•dis) is an opinion formed before all the facts are known. It can lead to stereotypes. A **stereotype** (STAYR•ee•oh•tighp) is an overly simple opinion or attitude about a group of people.

Stereotypes and prejudice can cause conflict because they are not based on reality. They may lead a person to say or do harmful things to another person for no good reason.

What are some causes of conflict between friends and family members?

▼ Understanding and accepting differences can help you avoid conflict with others.

Blend Images - KidStock/Getty Images

When Conflict Mounts

Sorry Time

Suggest that your family have *sorry time* once a week. Each family member thinks of something he or she did that week that he or she regrets. The person then apologizes. After apologizing, think about how you feel. Better? Worse? Why?

Unresolved conflict can cause stress. It can damage relationships. It's better to resolve the conflict and make peace. **Peace** is being free of unsettled conflict within yourself or with others. Unresolved conflict can also lead to violence. **Violence** is an act that harms yourself, others, or property. You can avoid some conflicts. You can keep other conflicts from getting to a point where someone uses violence. For example, if you feel that someone is angry with you, calmly ask the person what is wrong. The other person may have misunderstood something you said or did.

Remember to use I-messages and active listening skills. Sometimes a conflict builds because people do not understand each other. You may think that a friend disagreed with you. You worry about the disagreement. A conflict builds with your friend. But your friend may not disagree with you at all. Instead of worrying, try doing something to ease the conflict. Use good communication skills to talk about the situation.

There are other ways to keep a conflict from getting out of control. The chart below gives some ways to stop a conflict from building.

When Conflict Builds

Instead of	Try
letting a situation get too serious	using communication skills to keep the situation calm
punching someone	talking about the situation
hiding what you feel	expressing how you feel in a healthful way
teasing people who are different from you	getting to know them so that you can understand them better

How to Resolve Conflicts

When you face a conflict, the following steps can help you resolve it.

1. **Stay calm.** Take deep breaths. Relax.
2. **Talk about the conflict.** Tell the person that you want to discuss the problem. Use I-messages to express your feelings. Listen to the other person. Don't interrupt.
3. **List possible ways to settle the conflict.** Check out each way to settle the conflict. Use the *Guidelines for Making Responsible Decisions™*.
   - Is it healthful?
   - Is it safe?
   - Does it follow rules and laws?
   - Does it show respect for you and for others?
   - Does it follow your family's guidelines?
   - Does it show that you have good character?
4. **Agree on a responsible way to settle the conflict.** Decide which solution is best for everyone. Make sure it is a responsible solution. If you can't agree, walk away. Ask your parents or guardian or another responsible adult for help.

◀ **The best way to deal with conflict is to have open and honest communication.**

What can you do to resolve conflict if you feel that someone is angry with you?

When to Get Help

Write About It!

Write an Explanatory Statement Suppose Moises and Perry have a conflict over who should be captain of the baseball team. Explain how someone could mediate the situation. What questions would the mediator ask? How would he or she try to get Moises and Perry to agree on a solution?

Sometimes it is hard to arrive at a solution with which everyone agrees. Two people may be so angry that they have trouble talking to each other without fighting. When this happens, mediation might be helpful. **Mediation** (mee•dee•AY•shun) is intervention to resolve conflict.

A person who helps with mediation is a *mediator*. A parent or guardian or another responsible adult is the best choice for a mediator. The mediator helps the people in conflict find a responsible solution.

If you are involved in a conflict that requires mediation, be respectful. Work together. Be prepared to give and take until you reach a responsible solution.

▼ **A mediator must be a good communicator and problem-solver.**

Glow Images

How Mediators Mediate

A mediator must listen to all sides of a conflict. Then he or she can help resolve it. Here are five steps that mediators take to resolve conflict.

1. Ask each person to explain what caused the conflict. Then ask each person how he or she feels. Then have each person restate what he or she heard.
2. Ask each person what solution he or she wants. Restate what each person says.
3. Ask each person what he or she can do to resolve the problem. Each person involved may have to give and take a little to reach a responsible solution.
4. Ask each person to agree on a solution. Make sure the solution follows family guidelines.
5. Write out the agreement and have each person sign it.

What is a mediator?

ACTIVITY

LIFE SKILLS

CRITICAL THINKING

Resolve Conflicts

With a partner, think of a situation that could cause conflict. Role-play how you could resolve the conflict. Use the steps below.

1. **Stay calm.**
2. **Talk about the conflict.**
3. **List possible ways to settle the conflict.**
4. **Agree on a way to settle the conflict. You may need to ask a responsible adult for help.**

After you finish, role-play the situation again. How could a mediator help resolve the conflict?

LESSON REVIEW

Review Concepts

1. **Identify** causes of inner and interpersonal conflict.
2. **List** the steps to use in resolving conflict.
3. **Explain** how mediation works.

Critical Thinking

4. **Generalize** Explain why conflict happens even among close friends.
5. LIFE SKILLS **Resolve Conflicts** You are working on a group project in school. You disagree with another member of your group about who should write the report. What is the first step you should take to resolve the conflict?

LESSON 4

Health in the Family

You will learn . . .

- about different types of families and how to keep families strong.
- ways a family can influence the health of its members.

Vocabulary

- **family**, *A65*
- **cooperative**, *A65*
- **heredity**, *A66*
- **habit**, *A66*
- **lifestyle**, *A66*
- **environment**, *A66*

Some of your most important relationships are with members of your family. There are many kinds of families. Members of healthy families share love, caring, and respect.

Kinds of Families

A **family** is a group of people who are related in some way. Most people spend their lives in families, first as children and later as parents or guardians bringing up children. The following are some kinds of families.

- **A nuclear family** has a husband and a wife who raise one or more children.
- **A single-parent family** has one parent who raises one or more children.
- **A blended family** is formed when one or both parents have been married before. It may include children from either parent.
- **An extended family** includes other relatives, such as grandparents, aunts, and uncles, who might act as guardians in some cases.

Members of healthy families cooperate with one another. Being **cooperative** (koh•OP•ur•uh•tiv) means being willing to work together. It also means following guidelines set up by your parents or guardian. *Family guidelines* are rules that guide children to act in ways that protect their health and safety and develop good character.

Everyone in a family needs to follow rules and laws. There are rules and laws for how to cross streets safely, for example. These rules help keep all family members safe.

An extended family includes different generations of family members.

What are family guidelines?

ACTIVITY
BUILD Character
Private Journal

Trustworthiness Write an entry in your private journal. Describe how you felt before and after having an honest conversation with a family member. Being honest is sometimes hard. Have you ever broken something that belonged to a family member? Did you admit it right away, or were you tempted not to? If you are honest with family members, they will trust you. In writing your entry, tell how your honesty helped you and the other family member.

Family Influences on Health

ACTIVITY

Science LINK

Make a Poster

Choose one disease that can be inherited, such as diabetes, heart disease, or sickle-cell anemia. Find out how heredity affects a person's risk of the disease. What actions can help reduce those risks? Present a poster showing what you've learned to the class.

Members of the same family influence one another's health. Certain influences are inherited. **Heredity** (huh•RED•uh•tee) is the traits you get from your birth parents. You may see your heredity in your appearance and talents, for example. Members of some families share a higher risk for some diseases.

Your family also may influence your habits. A **habit** is your usual way of doing things. All your habits make up your **lifestyle**, or your way of living. You may learn habits by following the example of other family members.

Children have many physical, mental, emotional, family, and social needs. A healthful family environment helps meet these needs. **Environment** (en•VIGH•run•mint) is everything around you.

Most parents try to give their children an environment that meets their needs. They provide food, shelter, and health care. They praise and encourage their children. In healthy families, family members talk to each other about their thoughts and feelings. This helps their mental and emotional health.

How might taking time to talk benefit the health of family members?

Bill Branson/National Cancer Institute (NCI)

If a child's physical needs are not met, the child may not grow and develop normally. The child may not do as well in school or activities as other children.

If a child's emotional needs are not met, he or she may feel that no one cares. Children learn from family members how to have caring and respectful relationships with others. Emotional neglect may make it harder for a child to develop relationships in the future. He or she may have trouble expressing feelings and thoughts in healthful ways.

How does praise and encouragement from family members affect a person's emotional health?

Family Members	Physical Health	Mental and Emotional Health	Family and Social Health
Mom			

CRITICAL THINKING

Analyze What Influences Your Health

Some members of a family may serve as role models. Some may take care of a child's physical needs. Others may support and encourage a child. Make a chart showing how the members of a family influence health.

1. **Identify people and things that can influence your health.** Draw a chart with four columns, like the one on the left. Label the columns *Family Member*, *Physical Health*, *Mental and Emotional Health*, and *Family and Social Health*. Add a row for each family member.
2. **Evaluate how these people and things can affect your health.** How does each family member influence a child's health?
3. **Choose healthful influences.** What healthful habits could a child develop as a result of the positive influence of family members? Add them to the chart.
4. **Protect yourself against harmful influences.** Write how each family member helps protect a child against harmful influences.

LESSON REVIEW

Review Concepts

1. **List** four types of families and describe how they stay strong.
2. **Describe** how family members influence each other's health.
3. **Explain** what might happen to a child who is neglected.

Critical Thinking

4. **Predict** Explain what could happen if a young person did not follow family guidelines.
5. LIFE SKILLS **Analyze what Influences Your Health** During dinner your mother asks you how your day went. How does answering her question help improve your health?
6. LIFE SKILLS **Be a Health Advocate** How could you be a role model for younger family members?

LESSON 5

Facing Family Challenges

You will learn . . .

- how families share and cooperate.
- how family members face changes and challenges.
- how healthy families communicate.

Vocabulary

- adoption, *A70*
- foster child, *A70*
- separation, *A70*
- divorce, *A71*
- abuse, *A71*
- neglect, *A71*

Family members face many challenges together. They have responsibilities at home, at school, and at work. They may also have to cope with changes. There may be a new family member, an illness, a separation, or a divorce. When family members meet these challenges with caring and respect, they can help each other's mental and emotional health.

Family Balancing Act

Many families have to balance the demands of jobs, school, and home. It takes organization to meet these demands. The family needs to plan who will do what.

Parents and guardians may work to earn an income. At home they may cook, clean, and take care of the needs of other family members.

Children help the family, too. They do chores around the house to help their parents or guardians. This gets the work done sooner so families can have more leisure time together. It also helps children learn skills they will need when they grow up and have homes of their own.

As children grow, they take on more responsibility. Being responsible means that they do their chores as well as they can. They make sure they finish the chores they are asked to do. Their parents or guardians know that they can trust the children to do their part to keep the family working well.

Design a Calendar

Make a calendar for your family to use to plan a schedule. Schedules can help families plan how they will get chores done and how they will do all the activities they want to do. Decorate your calendar with pictures showing your family working together. Then work with your family to use it to help organize your time.

How do children help their families?

Family Chore Schedule

	Mom	Dad	Jesse	Grace
Every day	Fix dinner Drive children to school Make bed	Fix breakfast Pack lunches Make bed	Do the dishes Feed the dog Make bed	Set and clear the table Walk the dog Make bed
Once a week	Clean the bathroom Do laundry	Mow the lawn Clean the kitchen	Clean my room Vacuum the house	Clean my room Water the plants

▲ **Your responsibilities grow as you do. What do you do to help at home?**

Growth and Change

ACTIVITY
BUILD Character

Role-Play an Adoption

Caring Suppose a friend's parents just adopted a child. With a partner, role-play what you might say and do to help your friend manage his or her feelings. Remember to encourage your friend to talk to his or her parents or guardian.

Families grow and change over time. Sometimes these changes are difficult. Family members can work together to face the challenges they cause.

New Family Members

A child might be born or adopted. Through **adoption** (uh•DOP•shuhn), parents take a child of other parents into their family. Some families take in foster children. A **foster child** is a child who lives in a family without being related by birth or adoption.

When a new child enters the family, an older child may feel happy to have a new brother or sister. He or she may feel sad, lonely, or jealous because his or her parents need to spend time with the new brother or sister. Sharing feelings among family members can help children adjust.

▲ Some families take in foster children.

Illness

What happens if a family member becomes ill and needs care? Other family members can cooperate to help the person. They can work together to handle the chores the ill family member used to do. They can love and support each other through difficult times.

Separation and Divorce

Usually parents can solve problems in their marriage. If they cannot, they may decide to separate. A **separation** occurs when a couple is still married but the husband and wife are living apart. Sometimes they can solve their problems, and they decide to live together again.

Sometimes parents can't solve their problems. A **divorce** is a legal end to a marriage. Divorce can make children feel very sad even though they are not to blame. Talking helps children and their parents deal with the changes.

Abuse and Neglect

People should feel safe in their families. Unfortunately, this isn't always true. A family member may abuse or neglect other family members. To **abuse** someone is to treat him or her roughly or harshly. Yelling at or hitting family members is a kind of abuse. **Neglect** is the lack of attention or care. Not giving a child the food, shelter, or health care that he or she needs is neglect.

Family members may be injured by the abuse or neglect. They may feel afraid, sad, and lonely. They may be scared to tell anyone about the abuse or neglect. Their physical, mental, and emotional health may suffer.

Why do people harm members of their families? Sometimes abuse and neglect are caused by drug or alcohol abuse. Other times, it happens when family members feel stress and do not know how to express it in healthful ways.

Getting Help

A family in which abuse or neglect exists needs help right away. Families may also need help to adjust to changes. If you need help dealing with a family change, first talk to your parents, guardian, or another responsible adult. There are also resources in many communities to help families adjust to changes and challenges. School counselors, religious leaders, doctors, and social workers have special training to help families manage hard times.

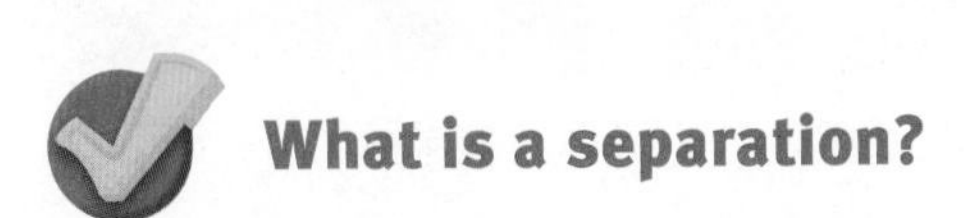

CAREERS

Sociologist

A sociologist is a person who studies the people and groups (like families) who make up society. A sociologist may look at attitudes, behaviors, and types of relationships. He or she may try to explain causes of social problems, including the problems some families face.

Family Ties

Write About It!

Write a Narrative Story
Write a story about a person who needs to communicate with family members about a problem. Write about why and how the family members communicate. What happens after they communicate? How do they solve the problem?

In a healthy family, family members feel a strong bond with one another. They can strengthen this bond by showing caring and respect for each other. They use words and actions to show how much they care. They know it is important to express their feelings daily. Here are some actions that help keep family bonds strong.

Make time for each other. Family members make time to talk to each other. They share thoughts and feelings. They care about what other family members have to say.

Show respect. Family members talk in ways that are open, honest, and positive. They don't put each other down. For example, they admit their mistakes and apologize for them.

Consider each other's feelings. Family members consider each other's feelings when they make decisions. Once parents or guardians make a decision, everyone honors that decision.

Share tasks. Family members share household duties. When everyone helps, work is done faster. This leaves more time for family members to do fun things together.

▶ **Keeping others informed is important to family communication.**

Resolving Conflict in Families

Family members can resolve conflict in healthful ways. They listen to each other. They cooperate to work things out. They apologize for problems they may have caused. They resolve conflict in ways that agree with family guidelines.

Family members may not always communicate in ways they know they should. Some disagreements take time to resolve. But even during rough times, family members still show care and respect for one another and remain loyal.

How do family members let each other know they care for one another?

LIFE SKILLS

CRITICAL THINKING

Access Health Facts

With a partner, make a booklet about one aspect of family health.

1. **Identify when you might need health facts.** Choose a subject related to family health that interests you. You might choose inherited diseases or ways to improve communication, for example.
2. **Identify where you might find health facts.** Make a list of sources you could check.
3. **Find the health facts you need.** Write down the health facts you find.
4. **Evaluate the health facts.** Decide whether your information came from valid sources.

Use the most reliable facts you found. Put them into a booklet. Add pictures to help explain what you found.

LESSON REVIEW

Review Concepts

1. **Describe** the benefits of sharing work in a family.
2. **Explain** changes and challenges a family might face.
3. **List** three ways families can show respect and caring for one another.

Critical Thinking

4. **Infer** Why is it important to get help right away if there is abuse or neglect in a family?
5. LIFE SKILLS **Access Health Services** How might a family find help in the community for a problem such as child abuse?
6. LIFE SKILLS **Practice Healthful Behaviors** Suppose a parent or guardian is ill. How might a student cooperate to help his or her family adjust to this challenge?

LESSON 6

Among Friends

You will learn . . .

- why friends in your neighborhood, school, and community are important.
- guidelines for strong friendships.
- how to avoid cliques.

Vocabulary

- **friend**, *A75*
- **bonding**, *A75*
- **social skills**, *A75*
- **clique**, *A78*

What makes a good friend? Good friends enjoy their time together. They encourage each other. They cooperate. They help each other to be the best they can be!

Friendship and Social Skills

A **friend** is a person who likes and supports you. You may be friendly with many people. You may develop a strong bond with one or two friends. **Bonding** is a process in which two people develop feelings of closeness for one another.

Friends spend time together, share mutual interests, and support each other when they face challenges. Friends can improve your social skills. **Social skills** are skills that help you interact with others. Knowing how to talk to other people in caring and respectful ways is one social skill. Being trustworthy is another.

You probably have many friends your age. You also may have friends who are teachers, coaches, or other responsible adults. These people may be part of your support network. These are people who can help you during hard times.

What is bonding?

▲ **What do you look for in a friend?**

Checklist for a Friend

- √ **Honesty** A friend tells you what's true, even when it's hard.
- √ **Trustworthiness** A friend won't tell your secrets.
- √ **Dependability** You can count on your friend to keep promises.
- √ **Fun** You enjoy doing activities together.
- √ **Politeness** A friend behaves in respectful ways.
- √ **Sincerity** A friend means what he or she says.
- √ **Patience** A friend doesn't give up easily.
- √ **Kindness** Your friend shows caring.
- √ **Respect** A friend doesn't tease you.
- √ **Family Guidelines** A friend doesn't ask you to do something of which your family disapproves.
- √ **Encouragement** Your friend wants you to do your best.

Making Friendships Stronger

Sometimes you will face challenges in friendships. Should you keep every secret that your friends tell you? How do you handle disagreements? You can help answer these hard questions with some guidelines for friendship.

- **Make responsible decisions with friends.** Use the *Guidelines for Making Responsible Decisions™*. Going along with decisions that you know are wrong can harm you. Wrong decisions also can cause you to lose the trust of your parents or guardian.
- **Keep confidences whenever possible.** Trust is an important part of friendship. You should usually keep your friends' secrets. However, you should tell an adult if the secret involves something illegal. You should also tell an adult if the secret could be harmful to your friend or to others.

ACTIVITY

Music LINK

Plan a Tribute

Plan a musical tribute for a special friend. Think about the qualities you like in your friend. You might think about your friend's kindness or your mutual interest in animals, for example. What songs make you think of these qualities? Record the parts of each song that remind you of your friend. Share your recording with your friend. If you can't make a recording, list the songs and share the list with your friend.

▶ **Friends show care and consideration for one another.**

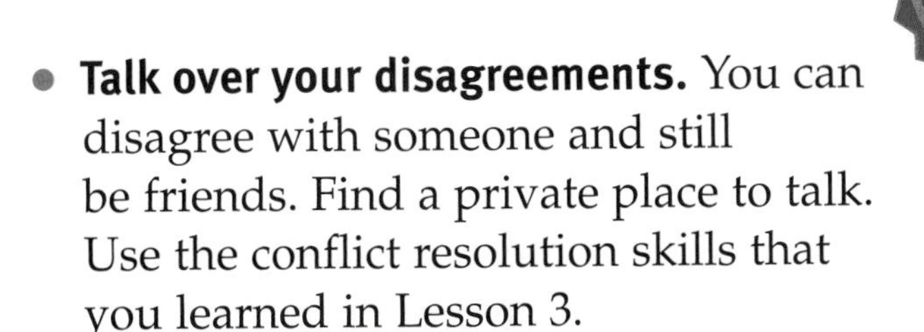

- **Talk over your disagreements.** You can disagree with someone and still be friends. Find a private place to talk. Use the conflict resolution skills that you learned in Lesson 3.
- **Keep your promises.** Think before you make a promise so you can be sure to keep it. Being dependable is very important in a friendship.
- **Encourage each other to grow.** When you are with a friend, think of new activities to enjoy together.
- **Allow each other time to be alone and to be with family members and other friends.** A healthful friendship should not limit you.

What can happen if you go along with a decision that you know is wrong?

ACTIVITY
LIFE SKILLS
CRITICAL THINKING

Set Health Goals

1. **Write the health goal you want to set.** "I will have healthful friendships." Choose one close friendship. Decide on a way to improve that friendship.
2. **Explain how your goal might affect your health.** How would working on your friendship affect your health? Record your thoughts.
3. **Describe a plan you will follow to reach your goal. Keep track of your progress.** Make a Health Behavior Contract. Plan to keep a private Friendship Journal for two weeks. Each night, write about what you did that day to strengthen your friendship. Did you spend extra time to listen to your friend talk about a problem he or she was having? Did you suggest a new activity to your friend? Record what you did and how it worked.
4. **Evaluate how your plan worked.** At the end of two weeks, evaluate your progress. Did you make an effort? Do you see results?

▼ What activities do you enjoy with friends? What new activities could you try together?

Randy Faris/SuperStock

Cliques Can Be Harmful

ACTIVITY

Health Online

Conflict occurs even among best friends. Think about a way you have resolved conflict with a close friend. Use the e-Journal writing tool to write a narrative describing the situation and what you did.

Is there a group of people you like to spend time with? Young people often form groups. People in the group may share common interests. These groups can help young people build social skills.

Sometimes a group can become a clique. A **clique** (CLEEK) is a small group of friends who stick together. The people in the clique may do everything together. They may tease or refuse to talk to people who are not in the clique. They may go out of their way to make outsiders feel left out.

Most people want to feel accepted and important. You may think that belonging to a clique will help you feel that way. But joining a clique may not be a healthful choice. Members of cliques hurt the feelings of people they leave out.

They can hurt themselves, too. People in a clique may limit the people they spend time with. They may not make friends outside the clique because they are afraid that the clique will reject them. They have less chance to learn social skills that help them get to know other people.

▼ **Accepting someone new into your group can help you make a new friend.**

Banana Stock/age fotostock

Avoiding Cliques

Social skills are important all through life. You can build your social skills and avoid cliques by using these tips.

- **Include new people** in your group of friends. As you grow up, you will meet new people. You can begin now to learn the skills you will need to form new relationships.
- **Go out of your way** to meet new people. You can learn a lot from people with different interests. Getting along with people who are different from you is one of the most important social skills you can develop.
- **Say "no"** if your group of friends wants to leave out others for no good reason. Tell them that you want to be friends with many different people.
- **Get together with others** in healthful ways. Throw a party, organize a study group, eat lunch together, or plan an event for a charity. These are all great ways to practice your social skills.

What can you do to build social skills?

LESSON REVIEW

Review Concepts

1. **Describe** how peer and adult friends can enhance the quality of your life.
2. **List** six things you can do to strengthen a friendship.
3. **Explain** why cliques can be harmful.

Critical Thinking

4. **Evaluate** If a friend who is very sad tells you in secret that she plans to run away from home, what would be a responsible thing to do?
5. LIFE SKILLS **Set Health Goals** In what ways might you achieve a health goal of including new people in your group of friends?
6. LIFE SKILLS **Make Responsible Decisions** You promise to help a friend after school. Later another friend invites you to go swimming after school. Which choice shows good character?

LESSON 7

Facing Challenges in Relationships

You will learn . . .

- how your peers can pressure you.
- how to use resistance skills to avoid risk behaviors.

Vocabulary

- **bully**, *A81*
- **chat room**, *A82*

Friends influence one another in many ways. You and your friends have fun together. You help shape each other's character. You support family guidelines. You help each other make healthful choices.

Peer Pressure

Peer pressure is the influence people your age have on you. They may pressure you to go along with their opinions and to behave in certain ways. *Positive peer pressure* is influence to do healthful things. One example might be a friend inviting you to exercise with her.

Sometimes friends can influence you in wrong ways. *Negative peer pressure* is pressure to do something wrong. A friend might suggest that you do dangerous stunts on a bicycle. It's important to recognize and resist negative peer pressure.

Parents and guardians set up family guidelines to keep children safe and healthy. They want children to follow their guidelines. They want them to avoid risk behaviors.

Bullies

Some peers use violence or threats to try to influence you. A **bully** is a person who likes to threaten and frighten others. If a person bullies you at school, report the behavior to a teacher or coach. Change your routes or the times that you do things. Stay with groups of friends.

Resisting Pressure

If you have a strong self-concept, it will be easier to resist pressure from peers. You learned in chapter 1 about some ways to handle peer pressure. Remember the following tips.

- **Match your actions** to your words. Show that you mean what you say when you say "no."
- **Use resistance skills** and the *Guidelines for Making Responsible Decisions*™ when a peer asks you to make a wrong decision. See the list at right for reasons for saying "no" to risk behaviors.
- **Suggest an alternative risk-free activity.**

What kind of peer pressure influences you in wrong ways?

10 Reasons to Avoid Risky Behaviors

1. I want to follow family guidelines.
2. I want to respect myself.
3. I want to respect others.
4. I want to have a good reputation.
5. I don't want to feel guilty.
6. I want to protect my health.
7. I want to protect my safety.
8. I want to follow laws.
9. I want to show good character.
10. I want to show good citizenship.

Negative Influences

ACTIVITY

On Your Own

FOR SCHOOL OR HOME

Interview a Family Member

Talk to an older family member or friend. Ask him or her about a time when he or she felt peer pressure. How did he or she handle the situation? Write and illustrate the story. Share it with other family members.

You've learned to look out for negative peer pressure. There are other influences to watch out for, too. Things you see and hear around you can influence your behavior. Some of these influences can be positive. Some can influence you to try harmful behaviors, though. You can avoid some of these influences.

Movies and TV

There are other kinds of pressure, too. Movies and TV shows often show risk behaviors such as smoking. They make these behaviors look fun. They don't show the harm risk behaviors can cause. Follow your family guidelines.

Chat Rooms

A **chat room** is a part of a computer system in which participants can have live discussions with one another. You may not know who these people are. It is best to avoid chat rooms or use only those that your parents or guardian approve.

A stranger might use the Internet to try to find out where you live or if you are home alone. Do not respond to e-mail from people you do not know.

LIFE SKILLS

ACTIVITY

CRITICAL THINKING

Make Responsible Decisions

You want to make friends with a new student. Your best friend says not to. What should you do? Draw a comic book to show how to decide.

1. **Identify your choices.** You might let your friend influence you not to make a new friend. You might choose to make a new friend anyway. What else could you do? Draw yourself in the comic book thinking about your choices.
2. **Evaluate each choice. Use the *Guidelines for Making Responsible Decisions™*.** In your comic book, show each choice with the answer to each question.
3. **Identify the responsible decision.** Check this out with your parent or a trusted adult. Draw yourself doing the most responsible thing.
4. **Evaluate your decision.** Draw a scene explaining your choice to your friend. What could you do to help your friend accept your decision?

Guidelines for Making Responsible Decisions™

- **Is it healthful?**
- **Is it safe?**
- **Does it follow rules and laws?**
- **Does it show respect for myself and others?**
- **Does it follow family guidelines?**
- **Does it show good character?**

What are two negative influences you can avoid?

LESSON REVIEW

Review Concepts

1. **Explain** what positive and negative peer pressure are.
2. **Describe** how you can resist when a friend applies negative peer pressure.

Critical Thinking

3. **Evaluate** How can you determine if peer pressure is positive or negative?
4. LIFE SKILLS **Make Responsible Decisions** You are on the computer. You notice a chat room that interests you. You don't have your parents' permission to use it. What would be the responsible decision?
5. LIFE SKILLS **Analyze What Influences Your Health** Give an example of how positive peer pressure can influence your health and behaviors.

Learning LIFE SKILLS

CRITICAL THINKING

Use Resistance Skills

Problem Your friend dares you and another friend to go into an old abandoned house. He says that there are fun places to play inside. Your parents have told you not to go in the house, and it also may not be safe. How could you say "no?"

Solution You can use resistance skills. Say "no" to an action that could be harmful to you or to others. It is not safe to go against family rules.

Learn This Life Skill

Think of other times you might need to use resistance skills. Follow these steps to help you resist negative peer pressure.

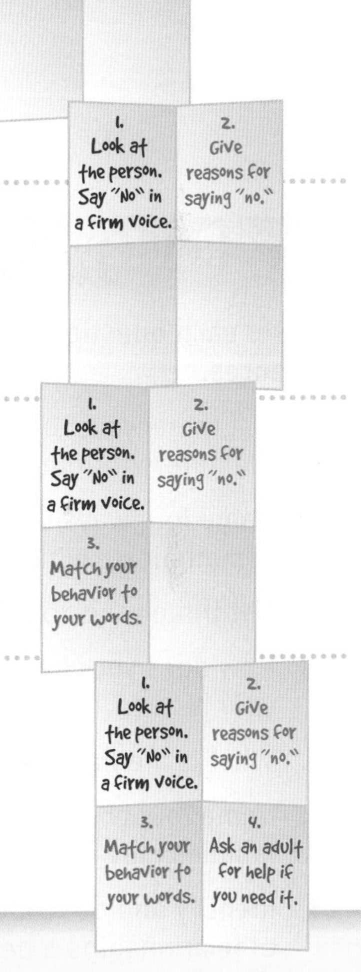

1 Look at the person. Say "no" in a firm voice.

You know that it is not safe to enter the building. It is also against your family guidelines. You know it is responsible to say "no."

2 Give reasons for saying "no."

Use one or two of the reasons from the list on page A81. Repeat the same reasons several times if necessary.

3 Match your behavior to your words.

If you think you will be pressured to go into the house again, stay away from the house when you are with your friend.

4 Ask an adult for help if you need it.

Tell your parents or guardian about the problem. They can help you find ways to resist this friend's pressure.

Practice This Life Skill

Suppose a friend asks you to play a game. Your parents have said you can't play until your homework is done. Your friend suggests that you tell your parents that you have no homework. With a group of classmates, role-play how to say "no" in this situation. How might you use the steps above?

CHAPTER 2 REVIEW

Use Vocabulary

body language, *A54*
clique, *A78*
neglect, *A71*
relationship, *A47*
role model, *A50*
separation, *A70*

Choose the correct term from the list to complete each sentence.

1. Parents may go through a __?__ before ending a marriage.
2. To fail to give a child proper care is __?__.
3. A small group of friends who stick together is a __?__.
4. You can be a __?__ for others by demonstrating healthful behavior.
5. Movements or gestures you make while talking are __?__.
6. A connection between you and another person is a __?__.

Review Concepts

Answer each question in complete sentences.

7. List three ways to make people who are different from you feel welcome.
8. Name four ways that people communicate with one another.
9. List three ways to use active listening.
10. How does your family influence your health?
11. What is mediation?
12. Name four changes or challenges that families may face.
13. Identify and describe the three types of relationships a person may have with others.
14. Describe four different types of families.

Reading Comprehension

Answer each question in complete sentences.

Peer pressure is the influence people your age have on you. They may pressure you to go along with their opinions and to behave in certain ways. Positive peer pressure is influence to do healthful things. One example might be a friend inviting you to exercise with her.

Sometimes friends can influence you in wrong ways. Negative peer pressure is pressure to do something wrong. A friend might suggest that you do dangerous stunts on a bicycle. It's important to recognize and resist negative peer pressure.

15. What is the difference between positive and negative peer pressure?
16. What is one example of positive peer pressure?
17. What is one example of negative peer pressure?

Critical Thinking/Problem Solving

Answer each question in complete sentences.

Analyze Concepts

18. You and a friend are having a disagreement about what movie to see. You are both starting to get angry. What should you do?

19. Explain how prejudice and stereotypes can lead to conflict and make it hard to get to know someone.

20. Your friend's mom just had a baby. What changes are they experiencing in their family? How can you help your friend adjust?

21. There is a new member joining your soccer team. How can you make her feel welcome and included?

22. You see two of your friends talking across the room. One of them is smiling and clapping her hands. The other has her arms folded across her chest and is looking at the floor. How do you think each person feels?

Practice Life Skills

23. **Make Responsible Decisions** You and Abby made plans a long time ago to go to the movies on Saturday. Friday night, Johnny calls. He has tickets to the circus for Saturday. What should you do? Use the *Guidelines for Making Responsible Decisions*™ to help you decide.

24. **Use Resistance Skills** You worked hard on your math homework last night. On the way to school this morning, a friend asks to copy all of your answers. How do you tell your friend "no"? Use resistance skills to plan what you would say.

Read Graphics

Joe is in fifth grade. He wants to use a chat room. He found out some information about several different chat rooms. He knows that his parents approve of a chat room on study tips.

Chat Rooms

Subject	Who Can Use It?
Study tips	Students in 3rd–6th grades
Rock bands	Anyone
Dating	Adults
Computer games	Students in 5th–12th grades

25. Which chat rooms might it be okay for Joe to use? Why?

26. Which chat room should Joe avoid? Why?

27. What should Joe do before visiting the computer games chat room?

UNIT A ACTIVITIES AND PROJECTS

Effective Communication

Make a Poster

Draw a poster with funny pictures showing Do's and Don'ts for active listening.

Self-Directed Learning

Perform a Skit

Visit the library to learn about ways to manage stress. With a group, write a skit as a fun way to give tips that you learned. Perform the skit for the class.

Critical Thinking and Problem Solving

Write a Story

Write a short story. The subject is a person who is very angry at a friend. The person is having trouble expressing his anger. Invent reasons for the anger. Why should the person express the anger? What are some ways to express it?

Responsible Citizenship

Conduct an Interview

Contact an organization in your community that helps families cope with their problems. Interview a staff member about what the organization does.

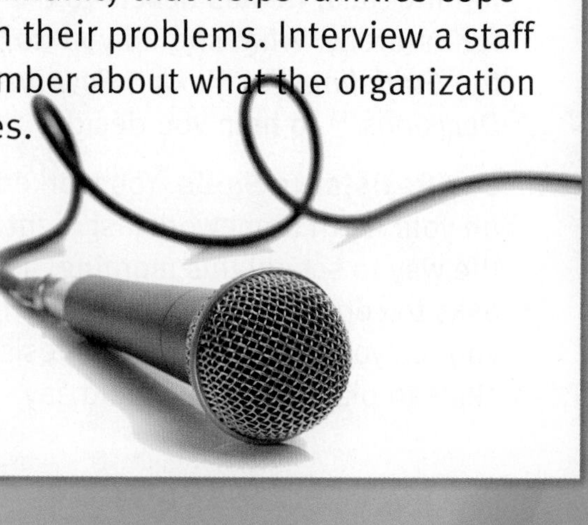

UNIT B

Growth and Nutrition

CHAPTER 3

Growth and Development, *B2*

CHAPTER 4

Nutrition, *B34*

CHAPTER 3

Growth and Development

Lesson 1 • Your Body's Systems B4

Lesson 2 • Your Heart and Lungs B10

Lesson 3 • More Body Systems B14

Lesson 4 • The Stages of Life B20

Lesson 5 • You Are Unique. B26

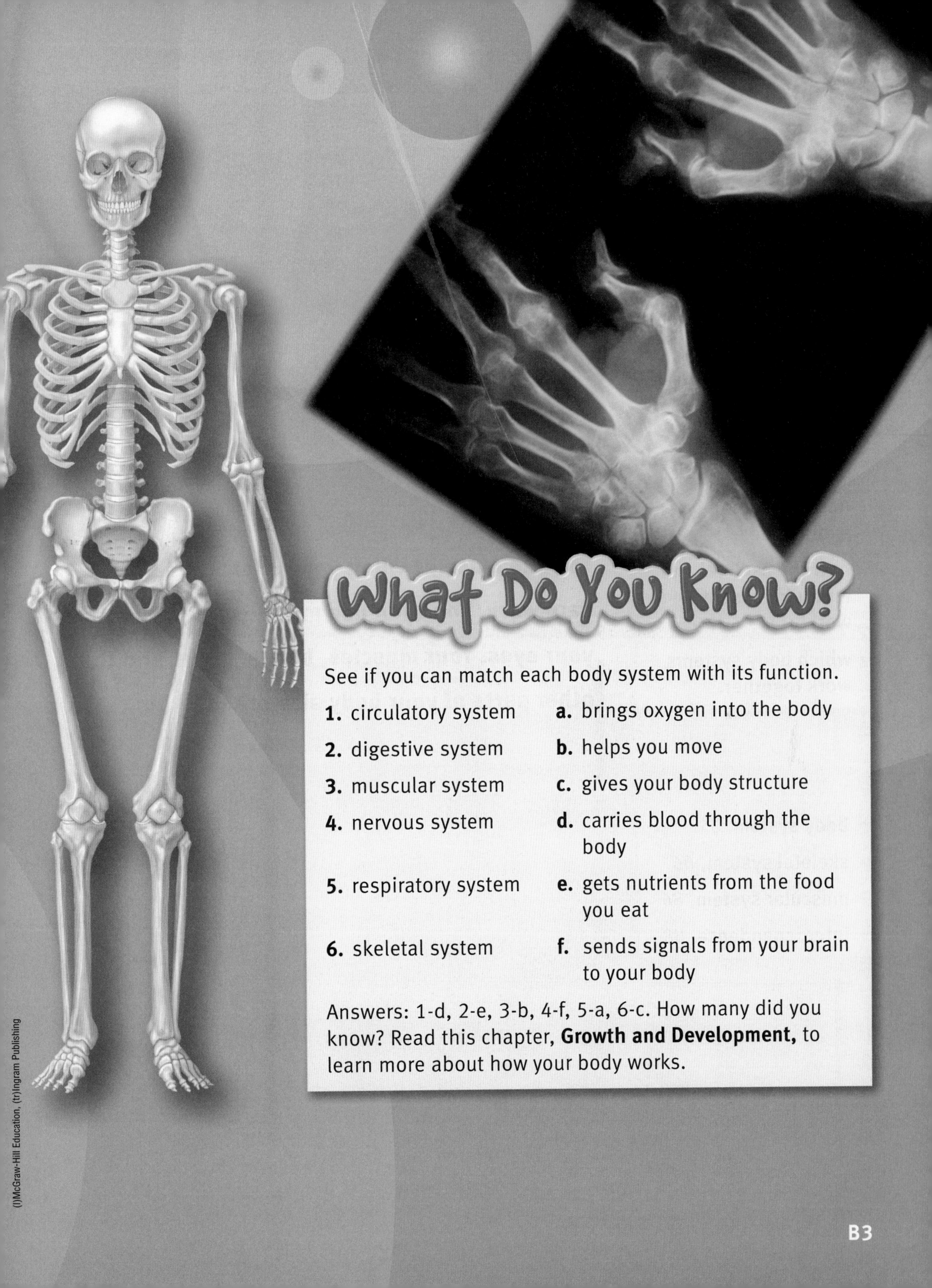

What Do You Know?

See if you can match each body system with its function.

1. circulatory system	**a.** brings oxygen into the body
2. digestive system	**b.** helps you move
3. muscular system	**c.** gives your body structure
4. nervous system	**d.** carries blood through the body
5. respiratory system	**e.** gets nutrients from the food you eat
6. skeletal system	**f.** sends signals from your brain to your body

Answers: 1-d, 2-e, 3-b, 4-f, 5-a, 6-c. How many did you know? Read this chapter, **Growth and Development,** to learn more about how your body works.

LESSON 1

Your Body's Systems

You will learn . . .

- how the human body is organized.
- how the skeletal and muscular systems work.
- which body systems work together.

Vocabulary

- **body system**, *B5*
- **skeletal system**, *B6*
- **muscular system**, *B6*
- **interdependence**, *B8*

You use muscles when you play sports and do hard work. You even use muscles when you blink your eyes. Your muscles, bones, brain, and the other parts of your body all work together.

Cells, Tissues, Organs, and Body Systems

The smallest living part of your body is a *cell*. Your body has many different kinds of cells. Blood cells, for example, carry oxygen throughout your body. Nerve cells carry messages to and from your brain. All cells use food and oxygen for energy. They grow and divide to form new cells. The new cells help you grow. They also replace cells that have died.

A group of cells that work together is a *tissue*. Your nerve cells make up tissues called nerves. Skin cells make skin tissue. Muscle cells make muscle tissue. All your body parts are made up of tissues.

A group of tissues that work together is an *organ*. Your heart, lungs, and kidneys are examples of organs. A **body system** is a group of organs that work together to carry out certain tasks. Your muscles all work together. They make up your muscular system. Your bones work together, too. They make up your skeletal system. Your heart and blood vessels work together. They make up your circulatory system. Each body system has a different function.

What is a tissue?

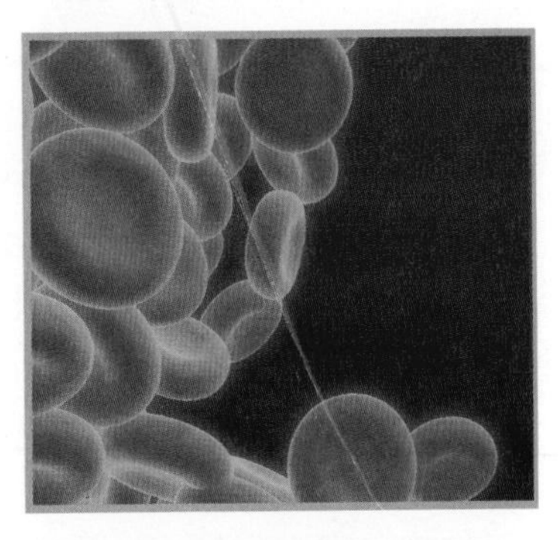

▲ **Cells are the smallest living parts of your body. These are blood cells.**

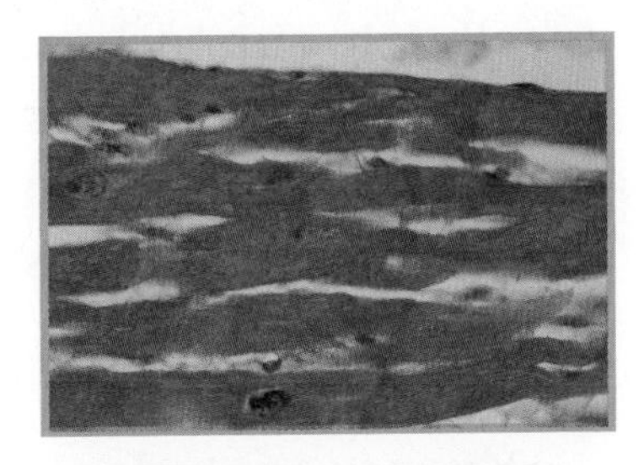

▲ **Cells make up tissues. This is muscle tissue.**

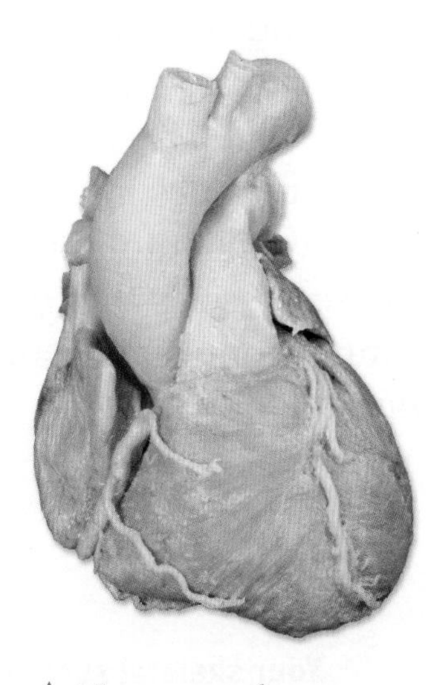

▲ **Tissues make up organs. Your heart is an organ.**

ACTIVITY

Science LINK

Investigate Replacement Parts

Find out which parts of the body can be replaced and how. Make a chart summarizing what you learn. Scientists and doctors can replace many body parts. People you know may have replacement hip bones, hands, or even legs.

Bones and Muscles

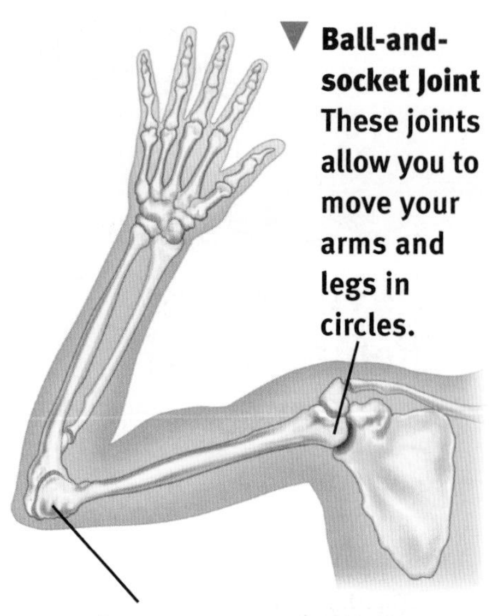

▼ **Ball-and-socket Joint** These joints allow you to move your arms and legs in circles.

▲ **Hinge Joint** Joints in your elbows and knees work like a door hinge.

Your body has more than 200 bones. They make up your skeletal system. The **skeletal system** is a framework that supports the body and helps protect internal tissues.

A *joint* is where two bones meet. Some joints move and some do not. Think about your elbow. Two bones meet there. The joint can move. Bands of tissue called *ligaments* (LIG•uh•muhnts) connect the bones in movable joints. Your skull is also made up of several bones. The joints between these bones cannot move.

The **muscular system** helps you move and maintain posture. Your bones and muscles work together to support and move your body. Muscles and bones connect to one another through bands of tissue called *tendons* (TEN•duhnz).

You can control some of your muscles, such as those in your arms and legs. Muscles that you can control are called *voluntary muscles.* Other muscles in your body work automatically. These muscles that you cannot control are called *involuntary muscles.* Your heart is an involuntary muscle.

The Skeletal System

skull
ribs
vertebrae
pelvic bone
femur

► **Your skeletal system lets you move and protects your internal organs. It gives your body shape and stability.**

Inside a Bone

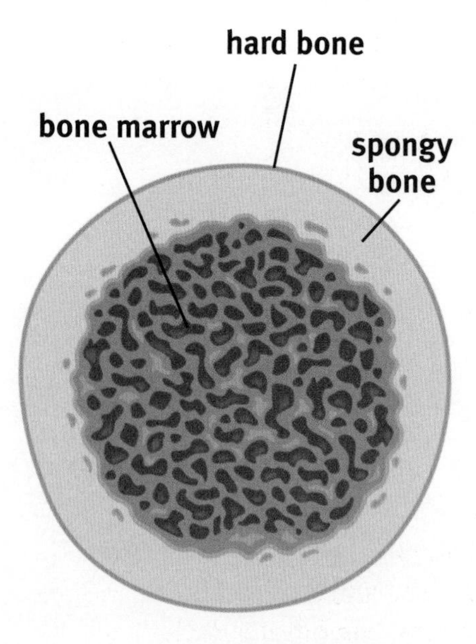

▲ **New blood cells are made inside bone marrow.**

Take Care of Your Bones and Muscles

Here are some tips to help you care for your bones and muscles.

- **Choose foods and drinks high in calcium.** Calcium helps make bones stronger. Milk, cheese, and juices with added calcium are good choices.
- **Get lots of physical activity.** Choose activities that build your muscles. You can swim, dance, jog, or ride your bicycle, for example. Physical activity that makes you support your weight can make your bones strong. Running and walking are good ways to do this.
- **Always use the right safety gear for the activities you do.** Helmets, wrist guards, and other safety gear protect your body.

Make an Exercise Poster

With a small group, choose one or two groups of muscles. Find out what kinds of exercise strengthen those muscles. Demonstrate the exercises for the class. Make a poster showing those exercises. Put your poster in your classroom or gym.

What is a joint?

The Muscular System

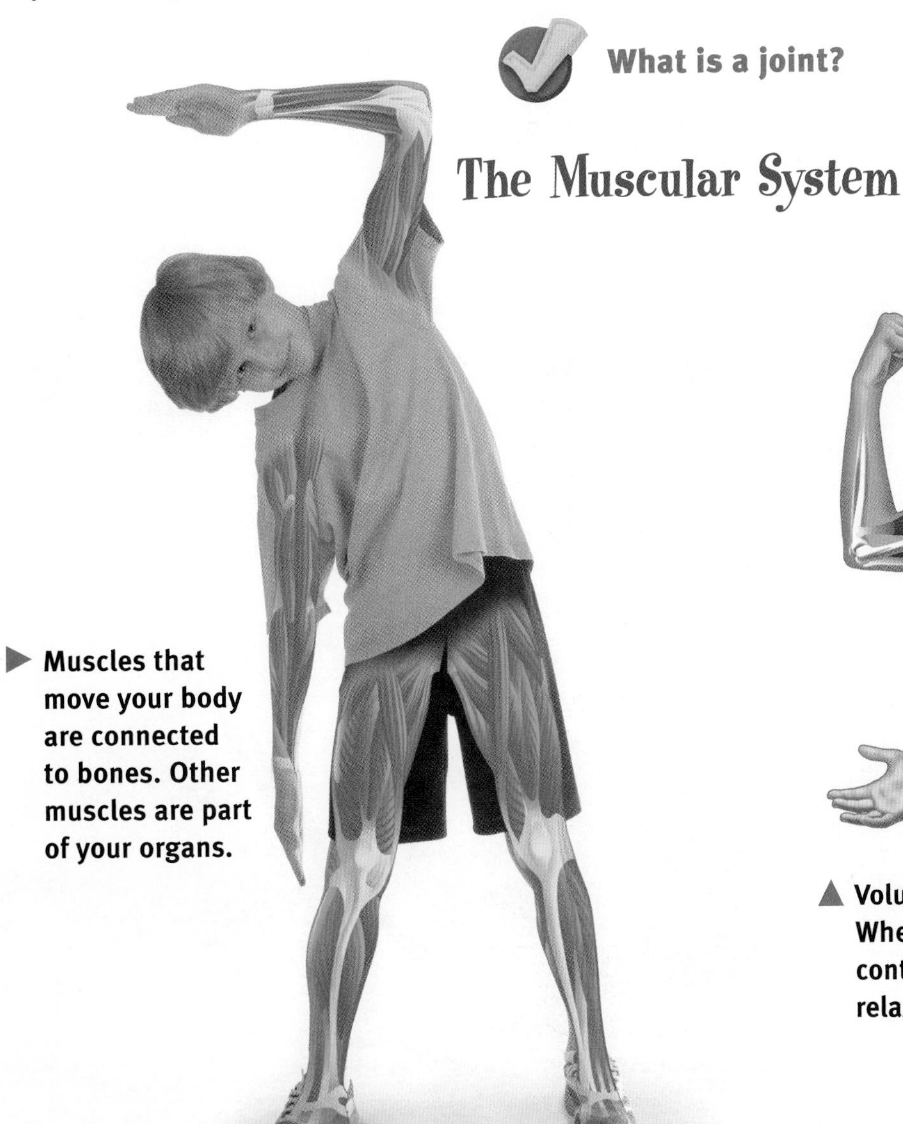

▶ **Muscles that move your body are connected to bones. Other muscles are part of your organs.**

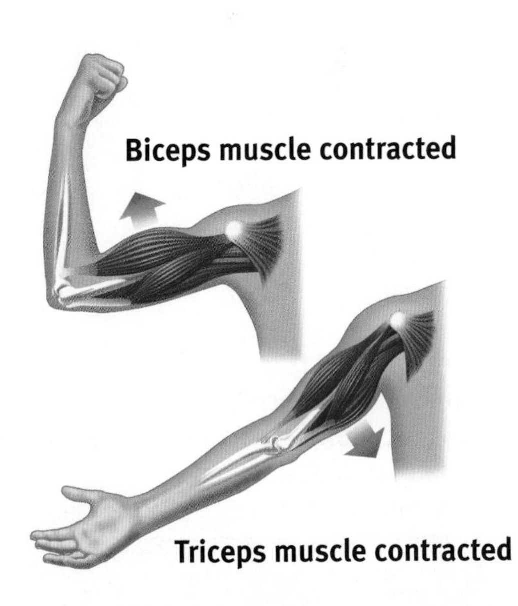

▲ **Voluntary muscles work in pairs. When one muscle in a pair contracts, the other muscle relaxes.**

Body Systems Working Together

Look at your hand. Make a fist. Now open your fist and spread your fingers. Many of your body systems had to work together to move your fingers. First your brain sent a message to your muscles. The message told them to move. Then the muscles pulled on tendons. The tendons pulled on the bones in your fingers.

This is one example of the interdependence of body systems. **Interdependence** means that body systems rely on one another to work properly.

CAREERS

Nurse Practitioner

Nurse practitioners are nurses who have special training in a particular area of health care. This allows them to do medical tasks that only doctors used to do. A nurse practitioner can give you a checkup, perform medical tests, and prescribe medicine. Nurse practitioners work in hospitals, clinics, doctors' offices, and schools.

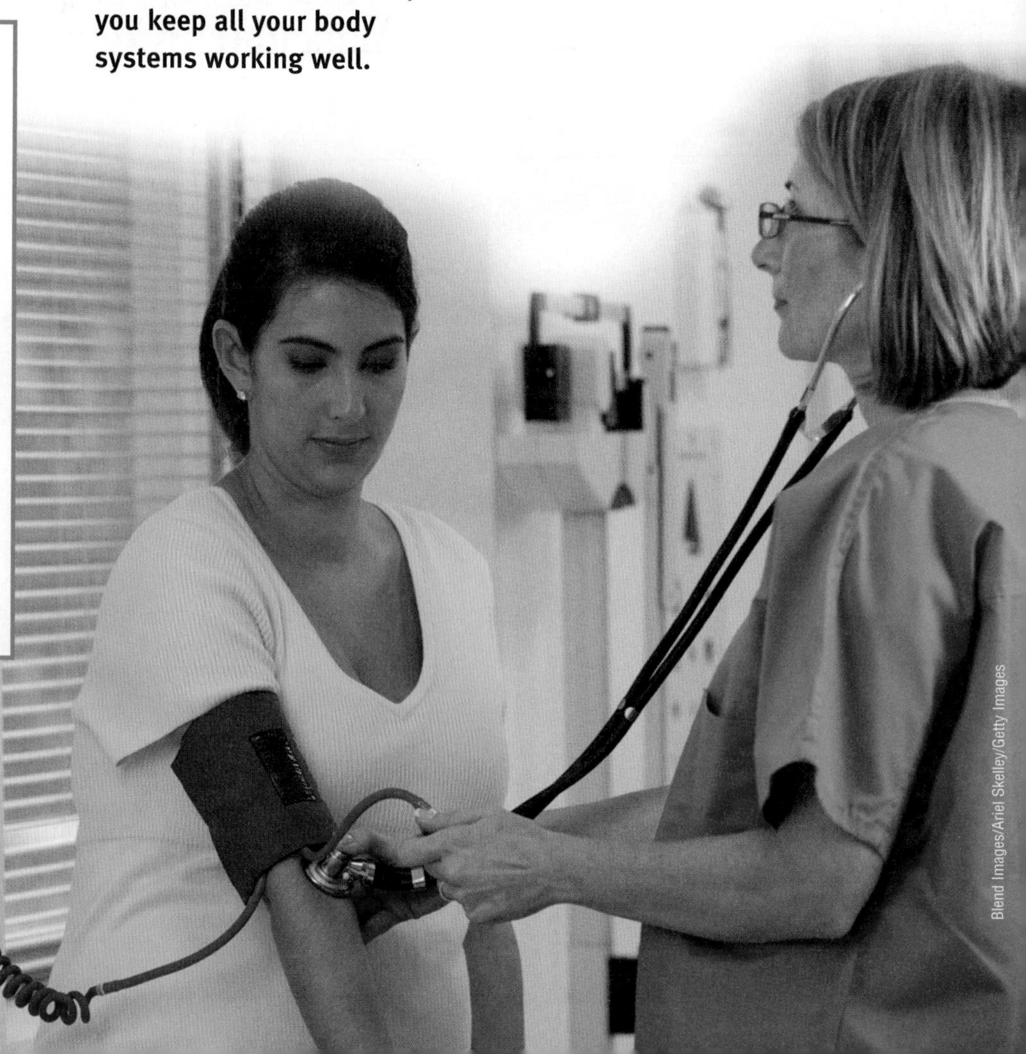

▼ **A doctor or nurse can help you keep all your body systems working well.**

What other body systems did you use to move your fingers? You needed energy to move them. Your respiratory system brought in air. You get oxygen from the air that you breathe. The oxygen helps break down food. Your digestive system changed food into energy. Your circulatory system carried energy through your blood to your bones and muscles. You'll learn more about all these body systems in the next few lessons.

A problem in one body system can cause problems in other body systems. For example, if the digestive system does not work well, other body systems won't get the energy they need. It's important to take care of all your body systems. Eating healthful foods and getting physical activity each day can help keep your body systems healthy.

Manage Stress

When you are stressed, your muscles may become tense. You can learn to relax your muscles. This helps protect your muscular system from stress.

1. **Identify the signs of stress.** Pay attention to your muscles. Do they feel tight?
2. **Identify the cause of stress.** What might be causing stress? Write down your ideas.
3. **Do something about the cause of stress.** Think about ways to solve a problem that is causing stress. List your ideas.
4. **Take action to reduce the harmful effects of stress.** Try to relax your muscles. Start at your toes and work up to your head. One by one, tighten each muscle. Then relax the muscle. Do you feel more relaxed? Write a short paragraph describing how your muscles felt before and after the exercise.

LESSON REVIEW

Review Concepts

1. **Describe** how cells, tissues, organs, and body systems are related.
2. **Identify** the functions of the skeletal and muscular systems.
3. **Describe** one way two or more body systems work together to keep your body working properly.

Critical Thinking

4. **Compare and Contrast** Explain the difference between voluntary and involuntary muscles.
5. LIFE SKILLS **Manage Stress** How can reducing stress protect your muscular and skeletal systems?

LESSON 2

Your Heart and Lungs

You will learn . . .

- how the circulatory system works.
- how the respiratory system works.

Vocabulary

- circulatory system, *B11*
- respiratory system, *B12*

Have you ever run for a few minutes? How did you feel? Did your heart pound in your chest? Did you breathe fast? Your heart and lungs worked together to help give you the energy to run. Together these body systems power your whole body.

Your Heart

Your **circulatory** (SUR•kyuh•luh•TAWR•ee) **system** transports oxygen, food, and waste through the body. Your heart, blood vessels, and blood make up your circulatory system.

Your heart is a muscle. It pumps blood to your body's cells through blood vessels. *Arteries* are blood vessels that carry blood away from the heart. Blood vessels that bring blood back to the heart are called *veins*. Arteries and veins end in tiny blood vessels called *capillaries* (KAP•uh•ler•eez). Blood moves, or circulates, through arteries to capillaries. Then it moves from capillaries to veins and back to the heart.

Here are some tips for caring for your circulatory system.

- **Get plenty of physical activity.** Like other muscles, your heart gets stronger when it works harder.
- **Limit fatty foods.** The fats can clog your arteries.
- **Have a plan to manage stress.** When you are stressed, your circulatory system works too hard. This can harm your heart and blood vessels.

What are the major parts of the circulatory system?

▶ **Blood vessels carry blood from your heart to all parts of your body and then back to your heart.**

The heart of an average adult pumps about 2,000 gallons of blood in one day.

Circulatory System

body cells

capillaries

▲ **Oxygen, nutrients, and wastes pass to and from cells through capillary walls.**

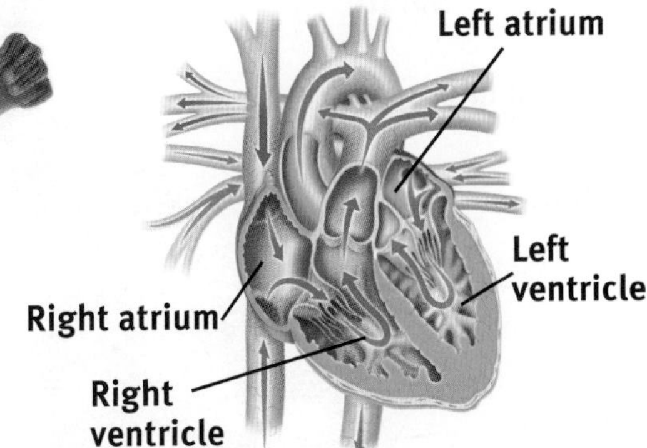

▲ **An atrium (AY•tree•uhm) is one of the two upper areas of the heart. Together they are called atria (AY•tree•uh). Ventricles (VEN•tri•kuhlz) are the two lower areas of the heart. The arrows show how blood moves in and out of the heart.**

Your Lungs

Write About It!

Write a Descriptive Poem List words related to the respiratory system, such as *breath, lungs, inhale, pant,* and *wheeze*. Use the words to write a poem about how it feels to take a deep breath.

Your **respiratory** (RES•puhr•uh•tawr•ee) **system** helps the body use the air you breathe. Air enters your body through your nose and mouth. It moves into your *trachea* (TRAY•kee•uh), or windpipe. From there it passes through your *bronchi* (BRAHNG•kigh), short tubes that carry air from your trachea to your lungs. The bronchi branch into many smaller tubes. These smaller tubes are called *bronchioles* (BRAHNG•kee•ohlz). Bronchioles end in alveoli (al•VEE•oh•ligh). *Alveoli* are small air sacs in your lungs.

The circulatory and respiratory systems work together. There are many capillaries that touch the alveoli. Oxygen moves from the air in the alveoli into the blood in the capillaries. Veins carry the blood to your heart. Your heart then pumps it to all the parts of your body.

As blood moves through your body, oxygen moves into the cells. At the same time, carbon dioxide moves from the cells into your blood. *Carbon dioxide* is a waste gas made by the cells. The blood carries this gas back to the lungs. The carbon dioxide moves from the capillaries into the alveoli. Then you breathe it out, or *exhale* it.

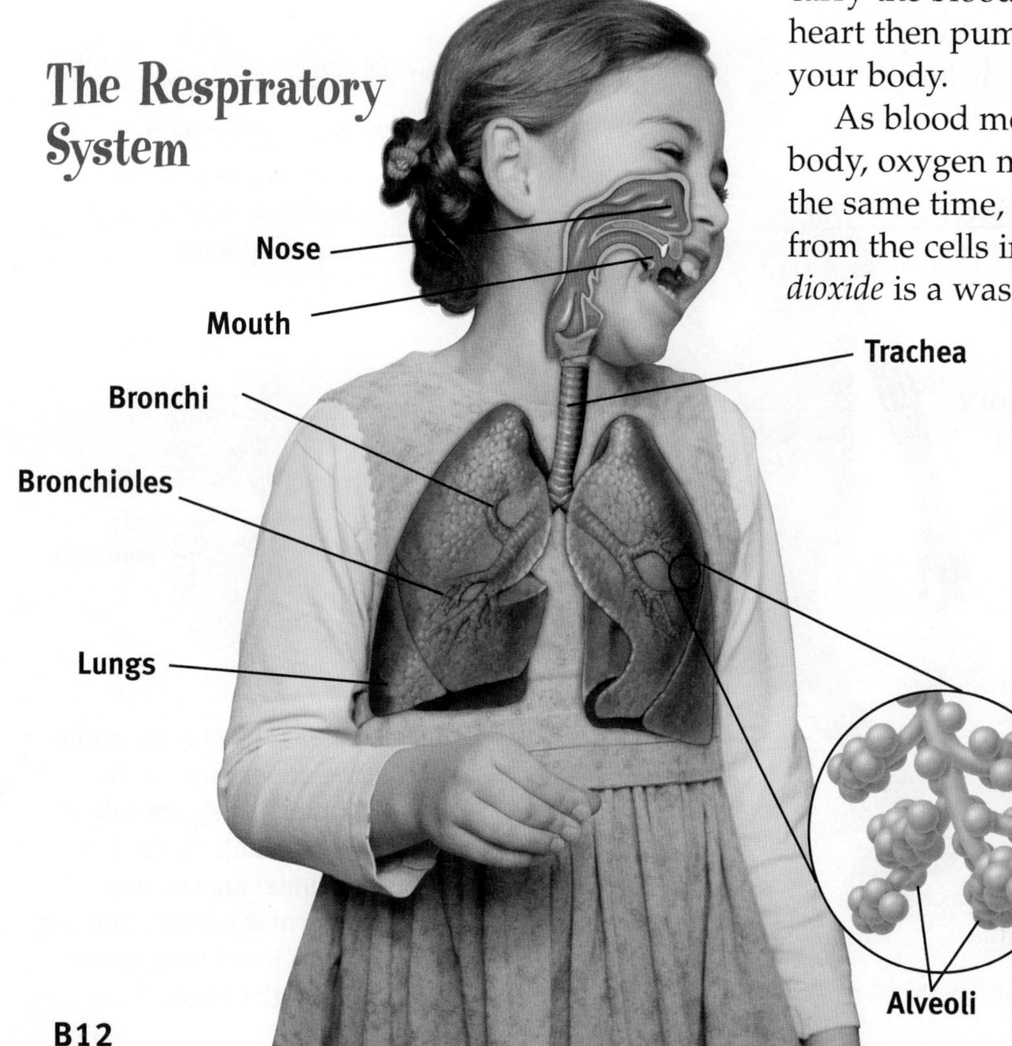

Take Care of Your Respiratory System

Here are a few good ways to care for your respiratory system.

- **Get plenty of physical activity.** When you are active, your lungs become stronger.
- **Avoid breathing poisonous fumes.** Fumes from cars, some cleaners, and some kinds of glue and paint can harm your lungs. Read the labels on glue, paint, and cleaners before you use them. Follow the safety instructions that come with the product.
- **Avoid smoking.** Tobacco smoke harms your lungs. It makes it harder for your lungs to move oxygen into your blood. Even breathing smoke from others can harm your lungs.

What are the parts of the lungs?

Be a Health Advocate

Make a poster to encourage people to keep the air around you clean.

1. **Choose a healthful action.** Keeping the air clean is a healthful action.
2. **Collect information about the action.** With a partner, brainstorm a list of ways that keeping the air clean helps your lungs.
3. **Decide how to communicate this information.** Design an eye-catching poster that tells why clean air is important for lungs.
4. **Communicate your message to others.** Put your poster in your classroom for your classmates to see.

LESSON REVIEW

Review Concepts

1. **Explain** how the circulatory system works.
2. **Explain** how the respiratory system works.

Critical Thinking

3. **Synthesize** Explain how the circulatory and respiratory systems work together to bring oxygen to your cells.
4. LIFE SKILLS **Practice Healthful Behaviors** What can you do to protect your circulatory system?
5. LIFE SKILLS **Be a Health Advocate** Physical activity helps keep both your circulatory and respiratory systems healthy. How can you encourage others to get more physical activity?

LESSON 3

More Body Systems

You will learn . . .

- how the digestive system works.
- how the nervous system works.
- how the endocrine and urinary systems work.

Vocabulary

- **digestive system**, *B15*
- **nutrient**, *B15*
- **nervous system**, *B16*
- **endocrine system**, *B18*
- **diabetes**, *B18*
- **puberty**, *B18*
- **urinary system**, *B19*

Have you ever wondered why infants drink only liquids? It's because their bodies cannot digest solid food yet. As you grow older, your body becomes able to break down food into a form it can use.

Digestive System

Your **digestive** (digh•JES•tiv) **system** breaks down food so that it can be used by your body. *Digestion* (digh•JES•chuhn) is the process your body uses to break down food.

The teeth chew food and mix it with saliva. This starts to break food down. The tongue then pushes food into the esophagus (i•SOF•uh•guhs). Muscles in the esophagus then push the food into the stomach. Digestive juices in the stomach break down the food even more.

Next the food moves into your small intestine. Capillaries in the intestine absorb nutrients. **Nutrients** (NEW•tree•uhnts) are substances in food that your body needs for energy, repairing itself, and growing. Your blood delivers the nutrients to your body's cells.

Caring for Your Digestive System

To get the energy you need, keep your digestive system in good shape.

- **Chew your food well** before you swallow it.
- **Eat foods that contain fiber,** such as grains and fruit. Fiber helps move food through your digestive system.
- **Get plenty of water** each day. Water helps food move through your digestive system. You can get water from other sources, including juices, vegetables, and fruit.
- **Take time to relax.** When you are stressed, your body may not digest food properly.

Where does digestion begin?

ACTIVITY

Consumer Wise

Analyze Ads

Advertisements sometimes try to make you think that you have a medical problem. They want you to try a particular medicine. The next time you are watching television with your parents or guardian, look for advertisements for medicines that affect the digestive system. How do the ads try to convince you that you need the medicine?

The Digestive System

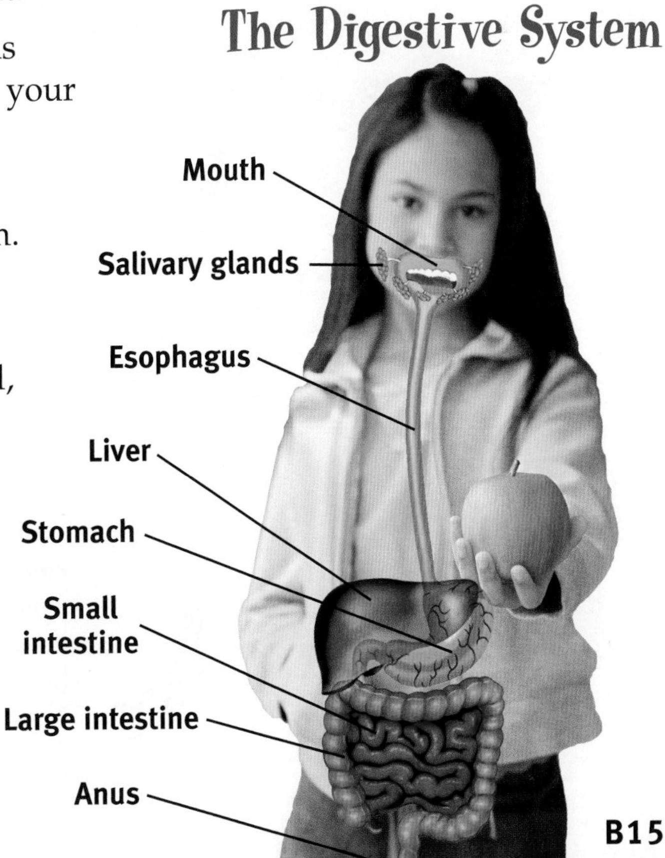

▶ **When you eat, food moves from your mouth through your digestive system. Undigested food leaves your body through the large intestine and the anus (AY•nuhs).**

ACTIVITY

Health Online

System Savvy

Choose a body system that interests you. Research that body system. Learn what organs and tissues it has, how it works, and how to care for it. Use the e-Journal writing tool to report on what you've learned.

Nervous System

Your **nervous** (NUR•vuhs) **system** controls all the functions of your body. It includes your brain, spinal cord, and nerves. Nerves are made up of nerve cells, or *neurons* (NEW•rahnz). Neurons carry messages between your brain and your body. They carry information very fast. It takes only a fraction of a second for a message to go from your hand to your brain.

Suppose that someone throws you a ball. Your eyes see the ball and send your brain a message. Your brain sends a message to your muscles. The message tells your muscles to move to catch the ball.

The Nervous System

Brain

Spinal cord

Nerves

Your Brain

Your brain has three major sections: the cerebrum, the cerebellum, and the brain stem. The *cerebrum* (suh•REE•bruhm) controls learning, memory, and voluntary movements. Each area of the cerebrum controls a different body function. For example, one area controls vision.

The *cerebellum* (ser•uh•BE•luhm) makes sure that your muscles work well together. It controls balance and keeps your movements smooth. Your *brain stem* controls actions such as your heartbeat, digestion, and breathing.

Parts of the Brain

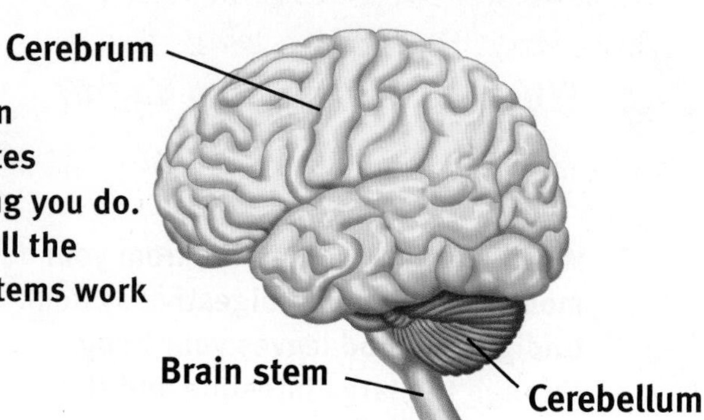

▶ **Your brain coordinates everything you do. It helps all the body systems work together.**

Your Spinal Cord

Your brain is the control center for your body. How does it control distant parts of your body like your hands and feet? Your *spinal* (SPIGH•nuhl) *cord* is a thick band of nerves that carries messages to and from the brain. Nerves branch off from your spinal cord to all parts of your body.

Your spinal cord sometimes acts as a control center, too. If you touch something very hot, you pull your hand away without even thinking about it. Nerves in the spinal cord tell the muscles in your hand to move—fast! This quick reaction without waiting for a message from the brain is called a *reflex*.

Caring for Your Nervous System

These tips can help you care for your nervous system.

- **Wear a seat belt** when you ride in a car. The seat belt keeps you from being thrown out of the car or into the windshield if there is a crash. It reduces the risk that you will injure your head or back.
- **Wear a helmet** when playing sports to protect your brain.
- **Avoid drugs,** including alcohol. Many drugs can cause brain damage.
- **Avoid breathing poisonous fumes,** such as those from glue or paint. They can harm your brain.

Access Health Products

Find out how to protect your nervous system when you play sports.

1. **Identify when you might need health products.** List some sports you play or would like to play. What protective equipment do you need?
2. **Identify where you might find health products.** Look in a phone directory to find a store that sells the equipment you need.
3. **Find the health products you need.** Call or visit the store to find out more about the equipment.
4. **Evaluate the health products.** Check the labels of the products you find. Do they meet government safety standards? Do they protect your nervous system? List the products you would choose to buy and explain why you would choose each one.

▲ **The products of technology, helmets and other equipment can protect your body systems when you play sports.**

How do the brain and spinal cord work together?

Endocrine and Urinary Systems

The pituitary gland is only about the size of a pea. Even so, it controls several other glands.

The **endocrine** (EN•duh•kruhn) **system** is made up of glands. The glands make chemicals called *hormones* (HAWR•mohnz). Hormones control specific body activities. For example, your pituitary (puh•TOO•uh•ter•ee) gland makes a growth hormone that controls how fast your body grows.

Another gland is the pancreas (PAN•kree•uhs). The pancreas makes a hormone called *insulin* (IN•suh•luhn). Insulin helps the body use sugar. A person may develop diabetes if his or her pancreas does not make enough insulin. **Diabetes** (digh•uh•BEE•tis) is a chronic disease in which there is too much sugar in a person's blood. You will learn more about diabetes in chapter 8.

The Endocrine System

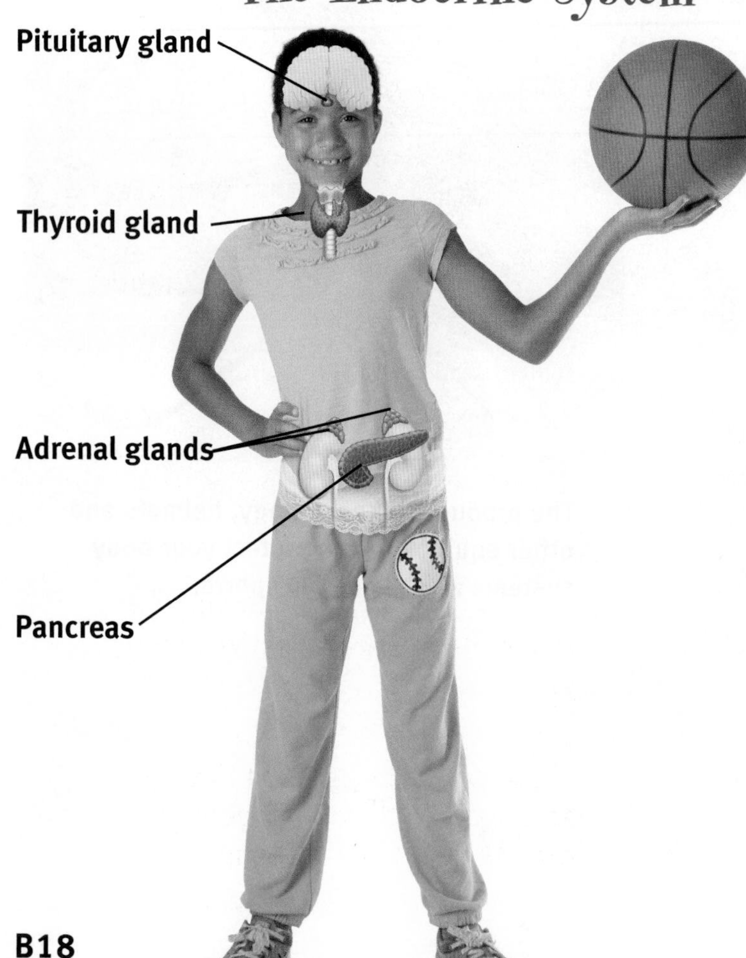

Hormones and Puberty

Puberty (PYEW•bur•tee) is the stage in life when a person's body becomes able to reproduce. The pituitary gland starts to make new hormones. These hormones affect the reproductive glands. In girls the reproductive glands are the *ovaries* (OH•vuh•reez). The reproductive glands in boys are the *testes* (TES•teez).

During puberty the ovaries and testes start to produce hormones. These hormones cause physical and emotional changes. You will read more about these changes in the next lesson.

◀ The hormones produced by the endocrine glands affect how you grow, change, and behave.

Urinary System

The **urinary** (YUR•uh•ner•ee) **system** removes liquid wastes from the blood. It also helps maintain the water level in your body.

Your cells produce wastes. These wastes are removed from the blood by the *kidneys* (KID•neez). The wastes combine with water to make urine (YUR•in). Urine passes from the kidneys to the bladder, which holds the urine until you urinate.

Here are some ways to take care of your urinary system.

- **Drink plenty of water.** This helps your kidneys to remove wastes.
- **Stay away from old paint.** Some old paint contains lead. Lead can harm many body systems, including your urinary system.
- **Take pain relievers only with the help of a responsible adult.** Taking too much of some pain relievers can harm your urinary system.

What are the main organs of the urinary system?

Kidneys

Ureters

Bladder

Urethra

ACTIVITY — On Your Own — FOR SCHOOL OR HOME

Drink More Water

You can keep your urinary system healthy by drinking plenty of water. Make a commercial encouraging people to drink the recommended six to eight glasses of water a day. Think of fun and humorous ways to get your point across. Write a script and perform your commercial for the class.

LESSON REVIEW

Review Concepts

1. **Describe** the structure and function of the digestive system.
2. **Describe** the structure and function of the nervous system.
3. **Describe** the structure and function of the endocrine and urinary systems.

Critical Thinking

4. **Analyze** Explain how drinking plenty of water affects your digestive and urinary systems.
5. LIFE SKILLS **Access Health Facts, Products, and Services** List three good sources of information about ways to care for your body systems.

LESSON 4

The Stages of Life

You will learn . . .

- what changes happen in infancy and childhood.
- what changes happen in adolescence and adulthood.
- how developing healthful habits helps you throughout your life.

Vocabulary

- **life cycle**, *B21*
- **growth spurt**, *B22*
- **critical thinking skills**, *B22*
- **mood swings**, *B22*
- **age**, *B24*

Life is a journey that you began as an infant. As you grow, you pass through different stages of life. Your body continues to develop and change with each stage. People develop at different rates. What stage of life are you in now? What stage will come next?

Infancy and Childhood

The **life cycle** consists of the stages of life from birth to death. At each stage, all parts of a person's health develop in certain ways. That is, a person grows physically, mentally, emotionally, and socially. The first stages in the life cycle are infancy and childhood.

Infancy *Infancy* runs from birth to about one year of age. Infancy is a time of rapid growth and change. Most infants begin to walk, talk, and eat solid foods during this year. Infants develop strong emotional bonds with their parents.

Early childhood *Early childhood* lasts from ages one to three. Children in this stage are called toddlers. They learn many new words. They also learn how to run, climb, and use their hands. Most toddlers learn to use the toilet, too. Children at this stage begin to develop relationships with other children. They develop ways of expressing happiness and sadness.

Middle childhood *Middle childhood* is the stage from three to six years. Physical growth is slower during this stage. Children grow socially, intellectually, and emotionally. They like to play with others. Children form friendships. They ask questions to learn more about the world.

Late childhood *Late childhood* begins at around age 6 and ends at around age 12. During this stage children develop their own interests and abilities. They can do more things on their own and are becoming more responsible. Spending time with friends becomes more important.

How is infancy different from childhood?

Adolescence and Adulthood

Changes During Puberty

- Your body becomes capable of reproducing.
- Hair grows in new places.
- Your perspiration may develop an odor.
- Pimples may appear on your face, neck, and back.
- Your critical thinking skills improve.
- You may feel strong emotions and mood swings.

Adolescence

Adolescence, the stage of life from 12 to 18 years, brings many changes. Puberty takes place during this stage. Puberty causes changes in all areas of your health.

Physical Changes The endocrine system produces hormones during puberty that cause physical changes. A child's body begins to look more like an adult's body. Most people have a growth spurt. A **growth spurt** is a rapid increase in height and weight.

During adolescence, sweat develops a stronger odor. Hair often grows in new places. The skin produces more oil. This can lead to skin problems. Pimples may grow on the face, neck, and back.

Mental and Emotional Changes Adolescents' critical thinking skills improve. **Critical thinking skills** are skills that help you think quickly and decide what to do.

The hormones produced during puberty can produce mood swings. **Mood swings** are rapid changes in emotions caused by hormones. Sometimes a person may go from being happy to upset in just a few minutes.

Social Changes Friends and social life are very important to most adolescents. Many adolescents have a few close friends with whom they spend time.

▶ **Adolescents learn new critical thinking and social skills.**

Adulthood

Like children, adults move through several stages. *Young adulthood* is the stage of life from 18 to 30 years. This is when many people learn how to live on their own. They have usually finished school and are finding work. They may marry and start families.

First adulthood is the stage of life from 30 to 45 years. People in this stage are often busy working at jobs and caring for families.

Second adulthood is the stage of life from 45 to 70 years. This stage is sometimes called midlife. By the end of second adulthood, some people have retired from their jobs. Their children are adults.

Late adulthood is the stage of life from age 70 on. Most people are healthy and active in late adulthood. Some have retired. People who retire may spend more time on other activities they enjoy. They may spend time with their families and share memories.

ACTIVITY

Social Studies LINK

Make a Presentation

Many cultures have traditions that celebrate a child growing into an adult. Choose a culture and find out more about its traditions. Give a presentation to your class about the tradition you chose.

▲ **Many adults raise families. What would you like to do as an adult?**

Death and Grief

Death is the final stage of the life cycle. When a person dies, his or her body organs stop working. When someone dies, you may feel grief. Grief (GREEF) is normal discomfort that results from a loss. A person feeling grief may not want to believe that a loved one has died. He or she may feel very angry or sad. He or she may have trouble sleeping or paying attention in school.

It's important to express grief in a healthful way. Write about your feelings. Talk to your parents or guardian or another responsible adult about what you feel. Keep your body healthy, too. Get enough physical activity and sleep. Eat healthful foods. If you feel very sad for a long time, tell your parents or guardian. A counselor may be able to help you.

List three physical changes during adolescence.

Healthful Habits for Life

ACTIVITY
BUILD
Character
Make an Album

Caring Collect or draw pictures of people you know from all stages of life. Below each picture, list a way you have shown that you care for that person. Arrange your pictures into a picture album. Share it with your family.

Life expectancy is the number of years a person can expect to live. In 1900 the life expectancy was around 40 years. Today most people can expect to live into their 70s. Choosing healthful habits now can help you age in a healthful way. To **age** is to grow older.

Here are six healthful habits you can start now.

- **Be physically active** to keep your heart, muscles, and bones strong all through your life.
- **Eat healthful foods** to give your body the nutrients and energy it needs.
- **Get enough rest and sleep** so that your body has time to grow and repair itself.
- **Keep your body clean** to reduce your risk of many diseases.
- **Protect your skin from the sun** and sunlamps. This reduces your risk of skin cancer later in life. Avoiding sun exposure also helps your skin look and feel healthy.
- **Avoid tobacco** to reduce your risk of lung and heart diseases later in life.

The habits you start now will help you stay healthy as you move from one stage of life to the next.

Why is it important to start healthful habits now?

Keeping your teeth clean will help them last all through your life.

ACTIVITY

LIFE SKILLS

CRITICAL THINKING

Make Responsible Decisions

A classmate has a tanning oil. She encourages you to use it to get a tan. She says, "You don't have to worry about skin cancer until you're older." What should you do? With two other people, write and perform a skit showing how to make the decision.

1. **Identify your choices. Check them out with your parent or trusted adult.** Make sure your skit includes more than one choice.
2. **Evaluate each choice. Use the *Guidelines for Making Responsible Decisions*™.** In the skit, include the answers to each question for each choice.
3. **Identify the responsible decision. Check this out with your parents or a trusted adult.** One of you will play a responsible adult. In your skit, show how you can work with a responsible adult to choose a responsible decision.
4. **Evaluate your decision.** In the skit, show what would happen if you made a responsible decision.

Rehearse your skit. Perform it for your classmates.

Guidelines for Making Responsible Decisions™

- **Is it healthful?**
- **Is it safe?**
- **Does it follow rules and laws?**
- **Does it show respect for myself and others?**
- **Does it follow family guidelines?**
- **Does it show good character?**

LESSON REVIEW

Review Concepts

1. **Describe** the stages of infancy and childhood.
2. **Describe** the changes that occur during adolescence and adulthood.
3. **List** healthful habits you can practice throughout your life.

Critical Thinking

4. **Compare and Contrast** How are a person in early childhood and a person in late childhood similar? How are they different?
5. LIFE SKILLS Make Responsible Decisions Why is it a responsible decision to avoid tobacco now? Show how you would use the *Guidelines for Making Responsible Decisions*™ to avoid tobacco.

LESSON 5

You Are Unique

You will learn . . .

- how you are unique.
- which factors affect development.
- how people learn.

Vocabulary

- **unique**, *B27*
- **experience**, *B28*
- **interest**, *B28*
- **learning disability**, *B29*

What do you enjoy doing? What do you do well? What have you learned? Your answers won't be the same as anyone else's. You are one of a kind. There's no one else just like you.

Heredity

You inherited many traits from your birth parents. The color of your skin, hair, and eyes are traits you inherited. The way your body grows and develops is another trait. Heredity is one reason people grow at different rates.

No one else has exactly the same set of traits that you do. This makes you unique. To be **unique** (yew•NEEK) is to be one of a kind. If you and a sibling have the same birth parents, you each received a unique set of traits from your mother and father. You may have some traits in common, but you will have some that are different. Only identical twins inherit exactly the same traits.

Write About It!

Write a Descriptive Essay
Choose someone you know. Write an essay describing the person. Describe what makes the person unique. Include physical traits. Also include the person's interests and experiences.

Give an example of an inherited trait.

▲ **These identical twins look similar. They have the same heredity. What makes them different?**

Draw a Self-Portrait

Use markers and paper to draw a self-portrait. Think about what makes you unique. Consider your heredity, experiences, personality, interests, and private thoughts. You might write a paragraph on the back that describes what makes you unique.

Other Factors

Identical twins receive the same set of traits from their birth parents, but each twin is still unique. Why is this? Factors other than heredity influence development.

Personality Your *personality* is the way you behave and feel. Some people are loud and talk to everyone. Others are quiet and like to sit alone.

Experiences An **experience** is something that you have done or seen. For example, maybe you moved to a new home when you were young.

Interests An **interest** is a desire to know about or to take part in something. Do you like to read about horses? Do you play a musical instrument?

Private thoughts Your *private thoughts* are those thoughts that you keep to yourself.

Learning Styles

Each person learns differently. This is part of what makes each person unique. Your *learning style* is the way you best gain skills and information. Some people learn best by reading. Others learn best by doing things with their hands. Others prefer to listen to another person telling them information. There are many ways to learn. How do you learn best?

However you learn best, you can improve your learning skills. The chart lists tips for using different learning styles.

Learning by . . .

Reading	Listening	Doing
• Try to remember key points. • Take notes. • Study the pictures or diagrams. • Look up words you do not know. • Discuss what you read.	• Pay attention to what you hear. • Sit or stand close enough to hear. • Ask questions.	• Pay close attention to step-by-step directions. • Watch someone use the steps before you try. • Practice several times.

Cooperative learning is learning and working on projects with other people. Cooperative learning gives you a chance to practice social skills and teamwork.

Someone in your group might have a **learning disability**, which is a condition that causes someone to have difficulty learning. There are different kinds of learning disabilities. Some people your age may not understand what they read. Others may have trouble paying attention for more than a short time.

People with learning disabilities may need special help or extra time with schoolwork. Many people with learning disabilities do very well in school. Knowing their learning styles helps people of all abilities gain new skills.

List three ways you can help yourself learn by listening.

ACTIVITY LIFE SKILLS **CRITICAL THINKING**

Analyze What Influences Your Health

What makes you unique? Create an equation that adds up to you.

1. **Identify people and things that can influence your health.** List one inherited trait, one part of your personality, one of your experiences, and one of your interests.
2. **Evaluate how these people and things can affect your health.** Below each trait you chose, write a sentence that describes how it makes you unique.
3. **Choose healthful influences.** Put plus signs between the traits. These traits and behaviors can influence you in healthful ways.
4. **Protect yourself against harmful influences.** Write down three things you can avoid that could harm your health. Put minus signs in front of them.

LESSON REVIEW

Review Concepts

1. **Tell** why you are unique.
2. **List** factors that affect your development.
3. **Explain** what a learning style is.

Critical Thinking

4. **Infer** What traits did you or someone you know inherit from birth parents? How do you know?
5. LIFE SKILLS **Analyze Influences on Health** Write a paragraph explaining how an experience or interest has affected you.

Learning LIFE SKILLS

CRITICAL THINKING

Practice Healthful Behaviors

Problem You want to choose a behavior that will help your muscular system. You decide to take a dance class. You are a little nervous about getting started. How can you help yourself make it a habit to go to class every week?

Solution You can use the steps on the next page to help yourself make a healthful behavior a habit.

Learn This Life Skill

Follow these steps to practice healthful behaviors. The Foldables™ can help you.

1 Learn about a healthful behavior.

How is dancing healthful? Dancing can help your muscular system. It strengthens your heart, leg, and arm muscles.

2 Practice the behavior.

How can you make sure you go to dance class every week? You can write down the time and day of the class on a calendar. What else can you do?

3 Ask for help if you need it.

Ask a parent or guardian to help you get ready for class. Invite a friend to join the class with you. How can the dance teacher help you?

4 Make the behavior a habit.

If you go every week, you've developed a new healthful habit. Keep track of how you feel in class. Can you do more complicated steps? Can you dance for a longer time without getting tired?

1. Learn about a healthful behavior.

1. Learn about a healthful behavior.
2. Practice the behavior.

1. Learn about a healthful behavior.
2. Practice the behavior.
3. Ask for help if you need it.

1. Learn about a healthful behavior.
2. Practice the behavior.
3. Ask for help if you need it.
4. Make the behavior a habit.

Practice This Life Skill

Choose another healthful behavior you can practice. With a group, make a booklet or brochure explaining the behavior and how to make it a habit. Be sure to include each of the steps listed above.

CHAPTER 3 REVIEW

Use Vocabulary

body system, *B5*
diabetes, *B18*
interdependence, *B8*
interest, *B28*
life cycle, *B21*
nutrient, *B15*
puberty, *B18*
unique, *B27*

Choose the correct term from the list to complete each sentence.

1. The stages of life from birth to death make up the __?__.
2. During adolescence, you will go through the period called __?__.
3. If something is one of a kind, it is __?__.
4. A desire to learn or know about something is a(n) __?__.
5. A(n) __?__ is a substance your body needs for energy, repairing itself, or growing.
6. A group of organs that work together to carry out certain tasks is a(n) __?__.
7. Body systems relying on one another to work properly is called __?__.
8. The disease in which there is too much sugar in a person's blood is __?__.

Review Concepts

Answer each question in complete sentences.

9. What is the role of the endocrine system?
10. What is infancy?
11. List the parts of the circulatory system.
12. Give an example of cooperative learning.
13. Describe how bones, tendons, muscles, and ligaments are related.
14. Describe how the human body is organized.

Reading Comprehension

Answer each question in complete sentences.

Your spinal cord sometimes acts as a control center. If you touch something very hot, you pull your hand away without even thinking about it. Nerves in the spinal cord tell the muscles in your hand to move—fast! This quick reaction without waiting for a message from the brain is called a *reflex*.

15. What is a reflex?
16. What tells your hand to move when you touch something very hot?
17. Why are reflexes important?

Critical Thinking/Problem Solving

Answer each question in complete sentences.

Analyze Concepts

18. Explain how the circulatory and respiratory systems are interdependent.

19. How do you learn best? What learning and study strategies do you use?

20. How are the joints in your skull similar to the joint in your elbow? How are they different?

21. You notice pimples on your skin. What might be happening? Why?

22. How can practicing healthful habits now help your health later in life?

23. In which stage of life are you now? Your parents? How can you tell?

Practice Life Skills

24. **Practice Healthful Behaviors** Explain why getting the equivalent of 6–8 glasses of water per day is a healthful behavior. How can you make it a habit?

25. **Make Responsible Decisions** A friend finds an old can of paint. He suggests that you use it to paint a fence. What should you do? Use the *Guidelines for Making Responsible Decisions*™.

Read Graphics

The chart shows some of the roles of your body systems.

System	Provide Energy or Nutrients	Movement	Remove Wastes	Control Body Functions
Circulatory	yes	no	yes	no
Digestive	yes	no	yes	no
Endocrine	no	no	no	yes
Muscular	no	yes	no	no
Nervous	no	yes	no	yes
Respiratory	yes	no	yes	no
Skeletal	no	yes	no	no
Urinary	no	no	yes	no

26. Which systems are involved in movement?

27. How are the respiratory and circulatory systems alike?

28. How are the circulatory, digestive, respiratory, and urinary systems alike?

29. How are the muscular and nervous systems different?

CHAPTER 4

Nutrition

Lesson 1 • Your Basic Nutritional Needs . . . *B36*

Lesson 2 • Aim for a Balanced Diet *B42*

Lesson 3 • Food That's Safe to Eat. *B50*

Lesson 4 • Your Weight Manager. *B56*

What Do You Know?

Read these statements. On a separate sheet of paper, write **T** for true statements and **F** for false statements.

__?__ **1.** A healthful diet involves eating the same foods every day.

__?__ **2.** The Nutrition Facts label on a food package tells you how much a food costs.

__?__ **3.** The kitchen counter is a safe place to thaw frozen food.

__?__ **4.** The fats you eat give you energy.

If you chose F for all the statements except the last one, you know some facts about nutrition. Read **Nutrition** to find out more about this important topic.

LESSON 1

Your Basic Nutritional Needs

You will learn . . .

- six kinds of nutrients and what foods provide them.
- how to use MyPlate to plan a balanced diet.
- the Dietary Guidelines.

Every day you make choices about what foods to eat. Food gives your body the nutrients it needs to work and play. Some food choices are more healthful than others. Choosing healthful foods helps you maintain good health.

Vocabulary

- **balanced diet**, *B38*
- **food group**, *B38*
- **MyPlate**, *B38*
- **serving size**, *B38*
- **Dietary Guidelines**, *B40*
- **calorie**, *B40*

Six Nutrients You Need

A healthful diet gives your body six kinds of nutrients. You need these nutrients every day.

Proteins (PROH•teenz) are used by the body to repair cells and to grow. Proteins also give you energy. Eggs, cheese, nuts, beans, meats, and fish are sources of protein.

Carbohydrates (kahr•boh•HIGH•drayts) are the body's main source of energy. You can get carbohydrates by eating whole-grain breads, cereals, rice, and pasta. Corn, carrots, and many other fruits and vegetables are also good sources of this nutrient.

Fats help your body store some vitamins. Fats also supply energy. Sources of fats include peanut butter, meats, salad dressing, and dairy products such as butter and cheese.

Vitamins help your body fight disease. Vitamins also help your body systems work well. There are many different vitamins. Each helps your body in a different way. Vitamin A helps your eyesight. Vitamin D helps your body use the mineral calcium. Vegetables, fruit, milk and other dairy products, meat, eggs, and grains all supply vitamins.

Minerals help your body work well and build new cells. A mineral called calcium (KAL•see•uhm) helps build bones and teeth. Iron helps your blood carry oxygen to your body. Minerals are found in meat, eggs, fruits, vegetables, milk, grains, and breads.

Water helps your body stay at the right temperature, digest food, and get rid of waste. You can get some of the water you need by drinking water as well as other fluids, such as milk and juice. You get some water when you eat fruits and vegetables, too.

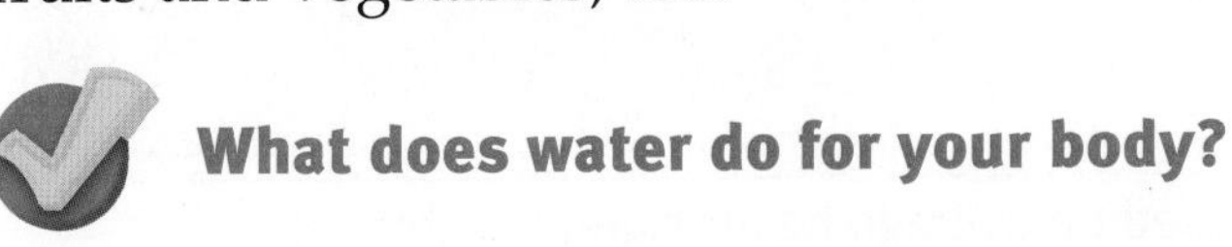

What does water do for your body?

MyPlate

Design MyPlate

On a piece of poster board, draw the outline of MyPlate. Fill in the plate with pictures of various foods from each food group. You can draw pictures or, with permission, cut out pictures from old magazines and newspapers.

A **balanced diet** is a daily eating plan that includes the correct amounts of food from the food groups. A **food group** is a group of foods that provide similar nutrients. The U.S. government created MyPlate to help you plan a balanced diet. **MyPlate** is a guide that shows what proportion of food is needed from each food group each day. A **serving size** is a specific amount or helping of food.

Find your balance between food and fun.

Move at least 30 minutes every day. Walk, dance, bike, rollerblade —it all counts!

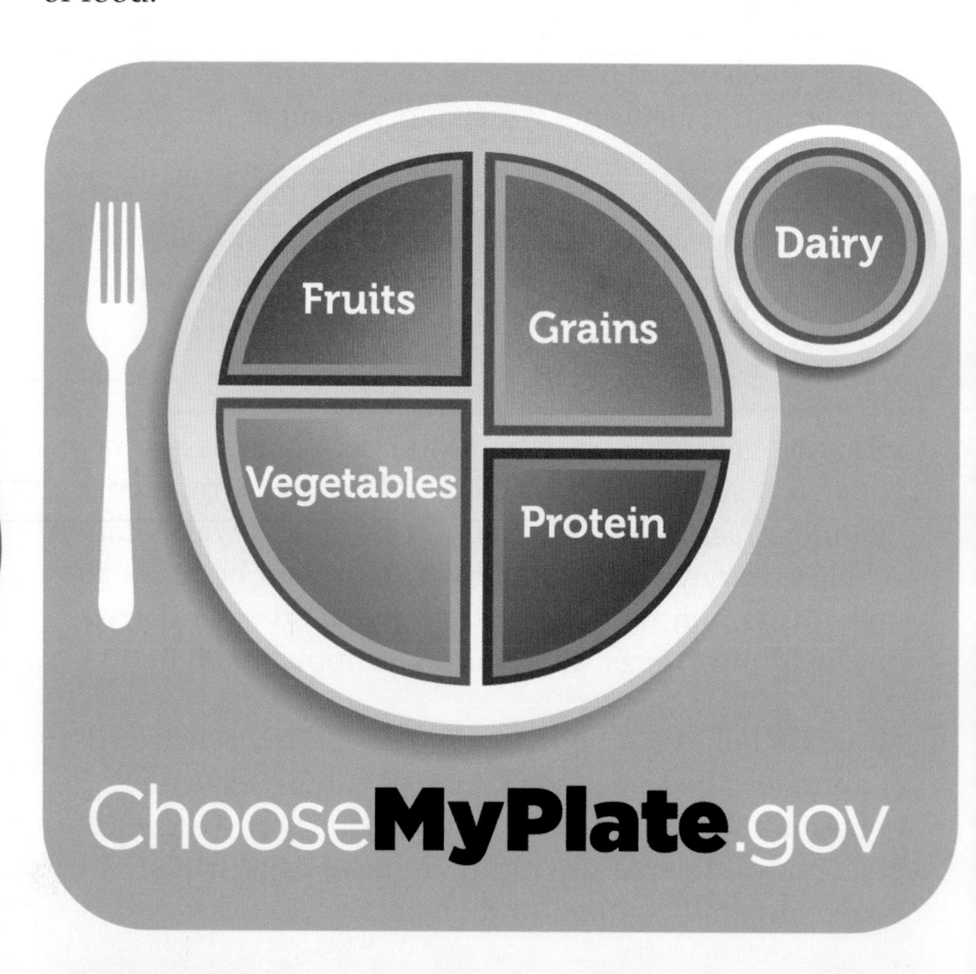

US Department of Agriculture

Oils

Oils are not a food group, but you need some for good health. Get your oils from fish, nuts, and liquid oils such as olive oil, corn oil, soybean oil, and canola oil.

Fats and Sugars

Get your facts from the Nutrition Label.

Limit sugars, fats, and salt. Choose foods low in added sugars and sweeteners.

Daily Food Plan	Grains	Vegetables	Fruits	Dairy	Protein
10-year-old girl who is active at least 30 minutes a day	5 oz.	2 cups	1.5 cups	3 cups	5 oz.
10-year-old boy who is active at least 30 minutes a day	6 oz.	2.5 cups	1.5 cups	3 cups	5 oz.

*Information is based on average heights and weights.

Grains Group This food group provides energy. Foods in this group are grains and foods made from grains. Grains supply carbohydrates, vitamins, and minerals. They also are a source of fiber. Fiber helps you digest food.

Vegetable Group Foods in this group supply vitamins, minerals, carbohydrates, and fiber.

Fruit Group Foods in this food group provide your body with vitamins, minerals, carbohydrates, and fiber.

Protein Group Foods in this food group are rich in protein. They also supply vitamins, minerals, and fats.

Dairy Group Foods in this food group are dairy products. These foods are sources of the minerals, calcium, and phosphorus (FAHS•fir•us). You need both of these minerals for strong bones and teeth.

Oils Oils are not a food group. These foods should be eaten in limited amounts.

List the food groups in MyPlate in order from the most to the fewest amounts needed each day.

ACTIVITY LIFE SKILLS **CRITICAL THINKING**

Access Health Facts

You can use the MyPlate to help you choose ethnic foods. Plan a meal at an ethnic restaurant.

1. **Identify when you might need health facts.** Suppose you are going to a Thai restaurant. You need facts about Thai foods.
2. **Identify where you might find health facts.** Your local library might be a good source of information.
3. **Find the health facts you need.** Look up information about Thai foods. Make a MyPlate chart. Draw or write Thai foods in each of the food groups.
4. **Evaluate the health facts.** Plan a meal of Thai foods you might order. Include servings from each food group in the chart.

The Dietary Guidelines

The **Dietary Guidelines** are suggested goals to help live a long and healthy life.

▲ Have a healthful snack when you are hungry. Drink plenty of water.

1. **Get the calories you need for your activity level, age, and gender.** The energy produced by food is measured in a unit called a **calorie.** Use MyPlate to help you choose healthful foods to get these calories.
2. **Aim for a healthful weight.** Find out what is the healthful weight for you. Staying at a healthful weight decreases your risk of high blood pressure, type 2 diabetes, and some cancers.
3. **Be physically active each day.** Physical activity burns energy that is also measured in calories. You will maintain your weight if you burn as many calories as you eat.
4. **Get the nutrients you need from healthful foods.** Eating more fruits, vegetables, whole grains and fat-free or low-fat milk and milk products helps to decrease the risk of cardiovascular disease and some cancers. It also helps to strengthen bones.
5. **Choose foods that are low in fat.** Saturated fats come from fatty meats and dairy products. You can get unsaturated fat from nuts, seeds, soybeans, and plant oils. Foods high in saturated fats and trans fats can increase the amount of cholesterol in your blood. High cholesterol increases the risk of heart disease. Unsaturated fats don't increase cholesterol. All fats are high in calories, though. The following are ways to choose foods low in fat.

▲ Playing group sports is one way to be physically active.

- **Eat grains, fruits, and vegetables.**
- **Choose fish, poultry, and lean meat.**
- **Avoid fried foods.**

6. **Choose foods and drinks with little sugar.** Sugary foods, soft drinks, or sports drinks have many calories but few vitamins and minerals. They also increase your risk of tooth decay.

7. **Choose and prepare foods with less salt.** Too much salt increases your risk of high blood pressure. Eat fresh foods instead of prepared foods. Many prepared foods have added salt. Eating foods that contain potassium, such as bananas, also helps reduce the effects of salt on blood pressure.

8. **Keep food safe to eat.** Germs can get into foods and make you ill. Wash your hands before you prepare and eat foods. Foods that should be kept cold should not be left out of the refrigerator for a long time. Make sure foods such as hamburgers are cooked thoroughly before you eat them.

9. **Do not drink alcohol.** Alcohol has many calories, but no vitamins or minerals. Drinking alcohol is illegal for anyone under the age of 21.

ACTIVITY

Consumer Wise

Read the Labels

List your three favorite foods. Bring in a food label from each of the foods you listed. With a partner, read the labels to find the total fat, sugar, and salt content for each food. Use what you learn to make a bar graph that shows the fat, sugar, and salt content for each food. Present your findings to your class.

What behaviors can help you limit the amount of fat, sugar, and salt in your diet?

LESSON REVIEW

Review Concepts

1. **Name** the six kinds of nutrients your body needs and two foods that provide each one.
2. **List** the amount you should eat from each food group in MyPlate.
3. **Outline** the Dietary Guidelines.

Critical Thinking

4. **Analyze** Identify three foods in your diet that are high in salt, sugar, and fat. Suggest more healthful foods you could eat instead. Explain why each choice is more healthful.
5. LIFE SKILLS **Access Health Facts** Make a grocery list of 25 foods and beverages. Check the Dietary Guidelines to see if the foods you listed promote good health.

LESSON 2

Aim for a Balanced Diet

You will learn . . .

- factors that influence your food choices.
- how to plan meals and snacks.
- what information is found on a food label.

Vocabulary

- **ethnic food**, *B43*
- **fast food**, *B45*
- **Nutrition Facts label**, *B46*

Think about what foods you ate yesterday. Why did you choose to eat each food? Many factors play a role in your food choices. To get what you need from the foods you eat, your choices should add up to a balanced diet.

Influences on Food Choices

Why do you choose the foods you do? It's important to know what influences your food choices. Then you can decide if the influence is healthful or not. The following factors can influence your food choices.

Personal Preferences You eat foods you like and avoid those you don't. Your preferences may change over time.

Family and Culture Your family may have food traditions. You may eat ethnic foods. An **ethnic food** is eaten by people of a specific culture. For example, curried vegetables are eaten in India, tortillas are Mexican, yucca and plantains are popular Latin American foods.

Peers You may eat some foods because your friends do. Your friends may influence you to eat the kinds of foods they do.

Emotions Some people lose their appetites or eat more when they are angry, lonely, sad, or tired. People may choose to eat foods that bring back happy memories.

Advertising Food and beverage makers try to get people to buy their products. They do this by spending a lot of money on ads. Ads don't always tell you whether a food or beverage helps you follow the Dietary Guidelines.

Cost and Availability Many foods and beverages are cheaper to buy when they are in season. Corn and strawberries, for example, are grown only at certain times of the year. Fresh orange juice also is seasonal. Store brands are often cheaper than name brand products.

Health Benefits You may choose foods and beverages because they are part of a balanced diet that follows the Dietary Guidelines.

How can emotions affect food choices?

Broadcast a News Report

Suppose that a frost kills orange trees in Florida. Companies that make orange juice will not have as many oranges to use. With a group, write a news report on how the cost of orange juice will change across the country. Make sure you explain how this might change people's food choices. Broadcast the news report to your class.

▲ Advertisements try to influence your food choices.

Plan a Menu

With your parent or guardian's help, develop a one-week menu of balanced meals. Use MyPlate to plan. Ask your parent or guardian if you can help prepare the meals on one day.

Meals and Snacks

Your body needs nutrients for energy, growth, and fighting disease. These nutrients come from a balanced diet.

Planning your daily meals and snacks can help you get the nutrients you need. How many meals and snacks do you have? You may eat breakfast, lunch, and dinner with snacks in between. You might eat five smaller meals with no snacks. These are both healthful choices.

Meal Planning

The table below lists the food groups from MyPlate. It also gives examples of amounts for each food group.

Think about what you eat each day. Do your meals match MyPlate and the Dietary Guidelines? Do you need to make changes? Suppose that you usually have a doughnut for breakfast. You could eat unsweetened cereal for this meal instead.

Serving Sizes for Different Foods

Food	Looks Like	Food	Looks Like
1/2 cup cooked cereal, rice, or pasta	ice cream scoop	2 ounces cheese	2 dominos
1 cup dry cereal	palm of hand	2 tablespoons peanut butter	table tennis ball
1 cup cut vegetables	fist	2–3 ounces meat	palm of adult hand
1 medium-sized piece of fruit	baseball	1 teaspoon butter	adult thumb

Combination Foods

To what food group does a peanut butter sandwich belong? Or a pizza? Or cereal with milk or fruit? These are examples of combination foods. *Combination foods* include foods from more than one food group. You can count the amounts from each food group that make up the food. For example, suppose you make a sandwich with two slices of whole-wheat bread, three ounces of sliced turkey, and two slices of Swiss cheese. It would contain

- 2 slices whole wheat bread is 2 oz. from the grains group, or 1/3 of the suggested amount.
- 3 oz. sliced turkey is just over half the daily suggested amount.
- 2 slices of swiss cheese (3/4 oz. each) counts as 1 cup of milk.

For more information on portions go to www.ChooseMyPlate.gov

Fast Foods

Many fast foods are combination foods. A **fast food** is a food that can be cooked easily and quickly, without much preparation, and then sold to customers.
Some fast foods are more healthful than others. To make healthful food choices, use the Dietary Guidelines.

What steps can you follow to plan meals and snacks for one day?

Plan to Share Food

Respect You invite a friend to your house for dinner next week. The friend is a vegetarian, but you are having chicken for dinner. Suggest a way that you and your friend could eat together even though you eat different foods.

What food groups are included in these combination foods?

Food Labels

The government requires that food labels be put on all packaged food. The front of the package tells you the name of the food and the amount of food in the package. The list of ingredients (in•GREE•dee•unts) tells you what's in a food. The **Nutrition Facts label** gives you information about the nutrients in the food. You can use the information on food labels to make healthful food choices.

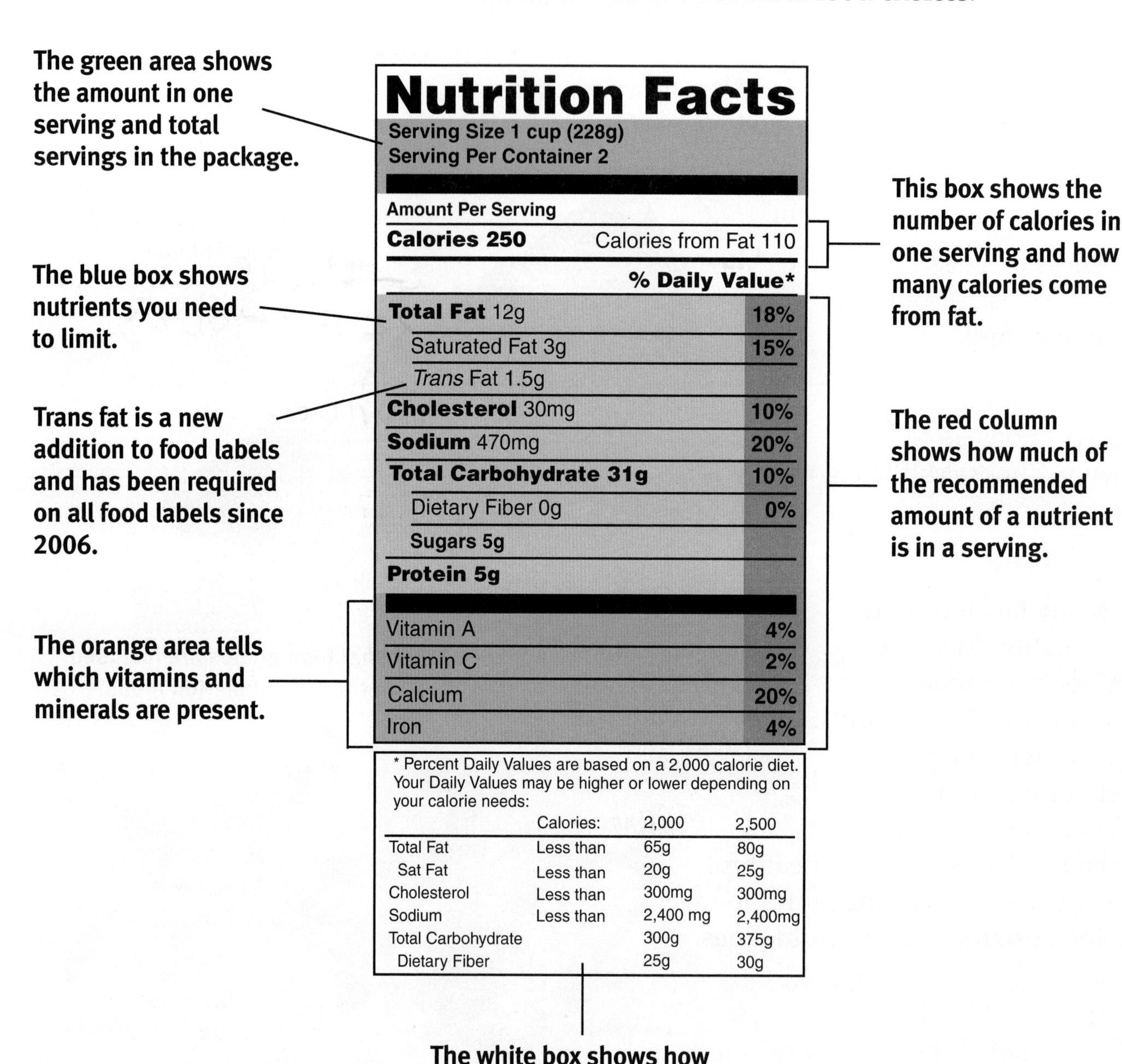

Nutrition Facts

Serving Size 1 cup (228g)
Serving Per Container 2

Amount Per Serving

Calories 250 Calories from Fat 110

	% Daily Value*
Total Fat 12g	**18%**
Saturated Fat 3g	**15%**
Trans Fat 1.5g	
Cholesterol 30mg	**10%**
Sodium 470mg	**20%**
Total Carbohydrate 31g	**10%**
Dietary Fiber 0g	**0%**
Sugars 5g	
Protein 5g	
Vitamin A	**4%**
Vitamin C	**2%**
Calcium	**20%**
Iron	**4%**

* Percent Daily Values are based on a 2,000 calorie diet. Your Daily Values may be higher or lower depending on your calorie needs:

	Calories:	2,000	2,500
Total Fat	Less than	65g	80g
Sat Fat	Less than	20g	25g
Cholesterol	Less than	300mg	300mg
Sodium	Less than	2,400 mg	2,400mg
Total Carbohydrate		300g	375g
Dietary Fiber		25g	30g

It is important to check labels before you buy food. You can compare the labels on two brands of cereal, for example. Check to see which has more of the nutrients you need.

Foods that are not sold in packages don't have a Nutrition Facts label. These foods include most fresh fruits and vegetables. Sometimes stores post nutrition information near these foods. You also can find information for these foods in a library or on the Internet. Often these foods are more healthful than packaged foods.

How can you use a food label to help you follow the Dietary Guidelines?

ACTIVITY

LIFE SKILLS

CRITICAL THINKING

Use Resistance Skills

Your friends want you to buy candy and soft drinks on the way home from school. You have set a health goal to eat healthful meals and snacks. With a friend, write a skit in which you resist the peer pressure using the steps listed below. Act out your skit for the class.

1. **Look at the person. Say "no" in a firm voice.**
2. **Give reasons for saying "no."**
3. **Match your behavior to your words.**
4. **Ask an adult for help if you need it.**

LESSON REVIEW

Review Concepts

1. **Name** seven factors that influence your food choices.
2. **Explain** how to plan healthful meals and snacks.
3. **Describe** the information you can get by reading a food label.

Critical Thinking

4. **Apply** Develop a healthful meal and snack plan for one day. Explain how each food or beverage you choose matches the goals in the Dietary Guidelines.
5. LIFE SKILLS **Use Resistance Skills** Your friends want to play video games. You have decided you want to be more physically active. List reasons you would say "no."

Learning **LIFE SKILLS**

CRITICAL THINKING

Analyze What Influences Your Health

Problem You are shopping for groceries with your parent or guardian. He or she asks you to choose a box of breakfast cereal. You reach for a box that is at your eye level on the shelf. It has a picture of your favorite sports star. Why was the cereal box placed on the shelf at eye level?

Solution

Many factors influence your choice of foods. Knowing what influences you can help you make responsible choices. Use the steps on the next page to check out these influences.

Learn This Life Skill

Follow these steps to help you check out influences on health.

1 Identify people and things that can influence your health.
The claims on food packages try to influence you to buy the food. The front of each box is an ad for the cereal. Look at the boxes of cereal in the picture. The packages have bright colors, catchy slogans, and special offers. These tricks try to make you think that the cereal will make you stronger, smarter, or more popular. How are ads for these three cereals trying to influence your choice?

2 Evaluate how these people and things can affect your health.
Look at each box again. Did the box give you information on how the cereal fits into the Dietary Guidelines and MyPlate? You can use this information to evaluate the cereal.

3 Choose healthful influences.
Which cereals would you buy? How did the box influence you to make a healthful choice?

4 Protect yourself against harmful influences.
Which boxes did not give you the facts you needed to make a healthful choice? How could you protect yourself from being influenced by a similar box?

	Analyze What Influences Your Health
Identify people and things that can influence your health.	The cereal box influences me...
Evaluate how these people and things can affect your health.	The box makes me think...
Choose healthful influences.	The box is a healthful influence because...
Protect yourself against harmful influences.	The second box is harmful because...

Practice This Life Skill

When you are in a grocery store, select a food with an attractive package. Use the four steps above to analyze the influence the package might have.

LESSON 3

Food That's Safe to Eat

You will learn . . .

- how pathogens can enter food.
- what safety guidelines apply to handling food.
- how to use table manners to stay safe and be polite.
- how to use the abdominal thrust to help someone who is choking.

Vocabulary

- **foodborne illness**, *B51*
- **pathogen**, *B51*
- **table manners**, *B54*

If food is not handled properly, pathogens can get into it. Eating this food can make you ill. Sometimes pathogens enter food before you buy it. Pathogens also can get into food if it isn't properly stored, prepared, or served.

How Pathogens Enter Food

A **foodborne illness** is an illness caused by eating food or drinking a beverage that contains pathogens. A **pathogen** is a germ that causes disease. Pathogens may enter foods and beverages at different times.

Growing and Processing

Some pathogens can enter foods and beverages before you buy them. The pathogens enter the food or beverage when it is being grown or processed.

- **Water that contains pathogens** can spread pathogens to plants or animals. Animals may drink the water. Plants can absorb it from rain.
- **Workers who do not wash their hands may touch foods.** Then pathogens enter food. If you eat the food, pathogens can enter your body.
- **Food may contact pathogens** on surfaces as it is processed. Suppose some meat has pathogens in it. A worker uses a machine to cut the meat. Then the worker uses the machine on another piece of meat. If the worker doesn't clean the machine first, pathogens can get into the new piece of meat.

Preparing, Storing, and Serving

Pathogens also can enter foods and beverages when they are prepared, stored, or served at home. Some foods and beverages need to be refrigerated or they may spoil. People may transfer pathogens from their hands to food. Dirty utensils can transmit pathogens between different kinds of food.

What causes foodborne illnesses?

CAREERS

Food Inspector

A food inspector works to keep food safe. Food inspectors examine animals and poultry to make sure that they are healthy. Food inspectors also visit processing plants to make sure that plant workers follow health rules. When inspectors find problems, they may tell the plant's owners to clean the plant or close it.

▲ **All meat and poultry must be examined by a food inspector before it can be sold.**

Safety Guidelines for Food

ACTIVITY

Health Online

Keeping It Safe

Choose a perishable food. A perishable food is one that can spoil if it is not kept at a safe temperature. Research how it is kept safe before it gets to your house. Also find out how to handle it safely at home. Use the e-Journal writing tool to write a summary of your research.

You can help protect yourself from foodborne illnesses. Use your eyes and your nose as you shop. Don't buy cans of food that are dented or bulging. Don't buy bruised fruit. Check the dates on food and beverage containers. Don't buy foods or beverages that are past the "sell by" date. Avoid discolored foods and beverages and those that smell bad.

You need to keep food safe after you buy it, too. There are four safety guidelines for handling food.

Wash Hands, Foods, and Utensils

Wash your hands with hot, soapy water before handling food. Afterward, wash counters, dishes, tools, cutting boards, and your hands with hot, soapy water. Wash raw fruits and vegetables before you eat them.

Store and Prepare Foods Separately

Store raw meat, eggs, poultry, and fish away from other foods. This can prevent pathogens from spreading from one food to another. Store raw meats, poultry, and fish in containers, so that they don't drip onto other foods. Wash all cooking utensils thoroughly after you use them with raw meat and before you use them with anything else.

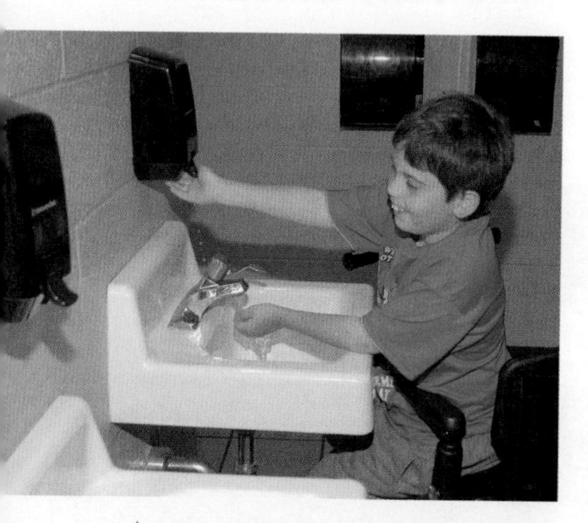

▲ **Wash hands often to prevent the spread of pathogens.**

▼ **Using a thermometer is the best way to make sure that meats are cooked completely.**

▲ **Prepare raw meats and vegetables on separate cutting boards to prevent the spread of germs.**

Cook Raw Food Properly

Follow safety guidelines when you cook food. Avoid food that contains raw or partially cooked eggs. Use a thermometer to make sure meat and poultry are cooked to the correct temperature. Cover and stir food that you cook in a microwave to be sure it cooks evenly. Covering the food keeps the heat in and helps kill some pathogens.

Keep Some Foods Cold

Many pathogens grow more slowly in cold temperatures. When you shop, leave frozen foods and other foods and beverages that need to stay cold, to pick up last. That way they won't be out of the freezer or refrigerator for too long. Most cold foods should be kept in the refrigerator at 40°F or lower. Put meat, poultry, fish, and eggs into the refrigerator or freezer within two hours after you buy them. If you freeze food, thaw it in the refrigerator, in a microwave, or in cold water. Change the water every 30 minutes. Put any leftovers away promptly. Use leftovers within three or four days.

MAKE a Difference

Ryan's Wells

Ryan Hreljac is an eleven-year-old from Ontario, Canada. Ryan has raised money for people in Africa who don't have clean water. The money will be used to dig wells to find clean water. Ryan's Well Foundation has raised more than half a million dollars. How does clean water help a person's health?

How does keeping food chilled or cold keep out pathogens?

▲ **Wrap and refrigerate leftovers right away.**

▼ **Always protect your hands when you carry hot food.**

Table Manners

Effects of Choking

Hold your finger near one end of a drinking straw. Then blow through the other end. You will probably feel the air blowing across your finger. Then make a small ball of clay. Put the clay into the straw. Then blow through it again. Did you feel any air? If food gets into your trachea, it can keep air from getting into your lungs just as the clay kept the air from passing through the straw. If you talk while you eat, food is more likely to enter the trachea.

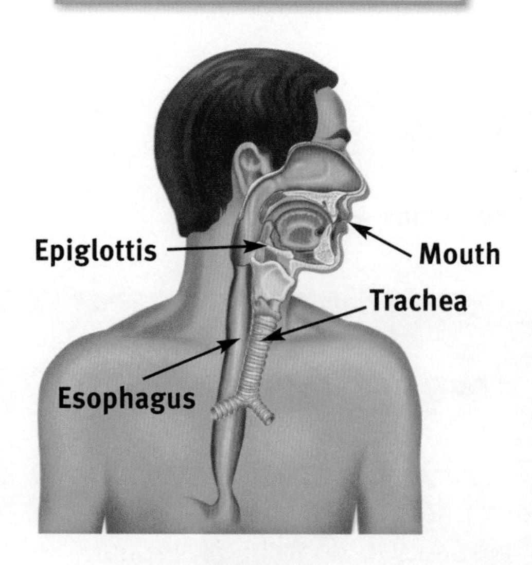

▲ **The epiglottis (eh•pi•GLAH•tis) keeps food from entering your trachea.**

Table manners are safe and polite ways to eat. They show respect for yourself and others.

Safety Benefits

Table manners keep you safe. Washing your hands before you eat reduces the risk of some illnesses. Using a napkin in your lap protects your clothing, the chair, and the floor from spills. Taking small bites and eating slowly reduces your risk of choking.

Social Benefits

Eating politely makes the meal more pleasant for everyone. Eating slowly in a pleasant environment aids digestion. Here are some manners to remember.

- **Set the table correctly.** This makes the table look nice and keeps dishes and silverware neat.
- **Wait until everyone is served** before you begin eating. You may start eating if someone who has not been served tells you that it is all right.
- **Share the food** with everyone at the meal. Take only your share.
- **Keep your mouth closed as you chew.** It's not pleasant to watch someone chewing with his or her mouth open.
- **Try all the foods served.** If you think you might not like one, take just a small amount. If you are allergic to a food, it is okay not to eat it.
- **Do not discuss stressful and unpleasant topics while eating.** Too much stress can cause indigestion. Keep the meal pleasant and relaxing.
- **Thank the person** or people who prepared and served the meal.

Help for Choking

A person who is choking may not be able to breathe. An abdominal (ab•DAH•mi•nul) thrust is one way to help someone who is choking.

- **Ask whether the person can speak.** A person who can speak can still breathe.
- **Take action** if the person can't breathe. Stand behind the person and wrap your arms around his or her waist.
- **Make a fist.** Place the thumb side of your fist just above the person's navel.
- **Grasp your fist** with your other hand. Quickly pull up and toward you.
- **Repeat** until the object pops out.

If you see someone choking, get an adult to perform the abdominal thrust if you can. You can do it if no one else is available. Never perform the thrust on a person who is not choking. You could harm their internal organs. You can practice with a partner how to place your hands.

What are the benefits of table manners?

LIFE SKILLS **CRITICAL THINKING**

Practice Healthful Behaviors

Set up a mock lunch in the classroom. Demonstrate healthful table manners that you could use at meals.

1. **Learn about a healthful behavior.** Before the lunch, work with a group to list some healthful table manners in a chart.
2. **Practice the behavior.** With your group, hold your mock lunch. Use healthful table manners. Record the manners you use in your chart.
3. **Ask for help if you need it.** Share your chart with your teacher and your parents or guardian.
4. **Make the behavior a habit.** For a week, continue recording what table manners you use at meals. Have you made a habit of practicing good table manners?

LESSON REVIEW

Review Concepts

1. **Explain** how pathogens can enter foods.
2. **List** the four safety guidelines for handling food. Give two examples of each.
3. **Describe** how to use table manners to stay safe and polite.
4. **Explain** how to perform an abdominal thrust.

Critical Thinking

5. **Analyze** Why is it important to keep hot foods hot and cold foods cold?
6. LIFE SKILLS **Practice Healthful Behaviors** Suppose you want to help make dinner. What can you do to help prepare the meal safely?

LESSON 4

Your Weight Manager

You will learn . . .

- how to maintain a healthful weight.
- how to find your healthful weight.
- the causes, signs, and treatment of eating disorders.

Vocabulary

- **healthful weight**, *B57*
- **weight management**, *B57*
- **overweight**, *B58*
- **underweight**, *B58*
- **body image**, *B60*
- **eating disorders**, *B61*

Maintaining a healthful weight helps you have energy. It also reduces your risk for some diseases. Maintaining a healthful weight also helps you look and feel your best.

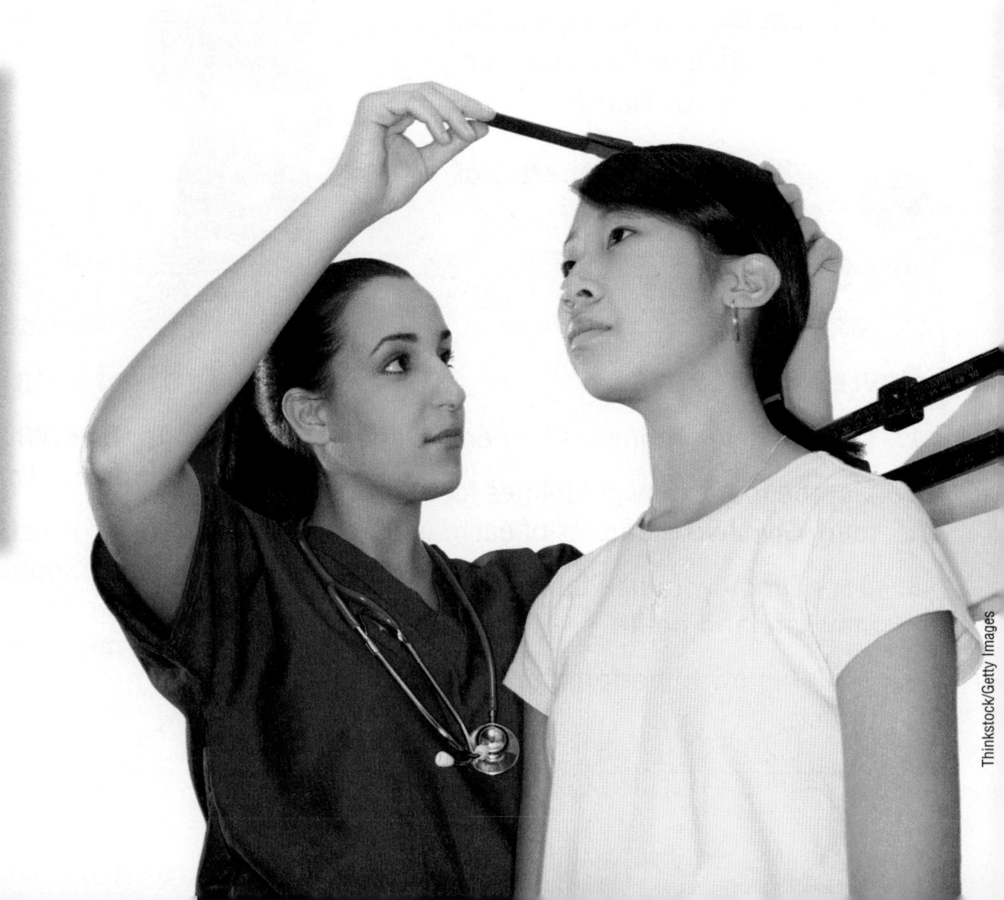

Thinkstock/Getty Images

A Healthful Weight

Your **healthful weight** is the best weight for your body. **Weight management** is a plan you use to have a healthful weight. Your weight stays about the same if you burn about the same number of calories as you eat. There are two parts to weight management. You need to get physical activity. You also need to eat the right amounts of healthful foods. Food labels usually say how many calories a serving or amount of the food contains. Food labels also tell you how much fat, salt, and sugar are in the food.You can use this information to plan healthful meals.

The chart below shows how many calories are in some common foods. It also shows how many calories are burned each hour by some kinds of physical activity.

Burning Calories

Duane weighs 100 pounds. He likes to ride his bicycle. After school each day, he rides at a speed of about 12 miles per hour for one hour. About how many calories does Duane burn in 5 days? About how many more calories will he burn if he rides at the same pace on Saturday and Sunday, too?

How does being physically active help you maintain a healthful weight?

Calories in Food			Calories Used by a 100-Pound Person	
Food	**Amount**	**Number of Calories**	**Activity**	**Calories Used Per Hour**
Broccoli (cooked)	1 cup	44	Swimming slowly	about 195
Bread (wheat)	1 slice	65	Bicycling (12 miles per hour)	about 290
Egg	1 large	75	Walking (3 miles per hour)	about 225
Blueberries	1 cup	81	Swimming fast	about 350
Low-fat milk	1 cup	102	Jumping rope	about 525
Skinless chicken breast (roasted)	3 ounces	142	Running (10 miles per hour)	about 900

▲ **To maintain a healthful weight, use about as many calories as you take in. The table shows how many calories a person who weighs 100 pounds would use. A person who weighs less would use fewer calories. A person who weighs more would use more calories.**

Weighing In

What happens if a person weighs more or less than a healthful weight? **Overweight** is weighing more than your healthful weight. Eating more calories than you use leads to being overweight. Being overweight increases your risk of heart disease, diabetes, and some kinds of cancer.

Underweight is weighing less than your healthful weight. Eating fewer calories than you use can lead to being underweight. An underweight person might be malnourished. This means not getting the nutrients needed for good health. As a result, the body might not be able to fight pathogens. Being underweight also can cause heart problems.

Write About It!

Write a Dialogue Write a dialogue between two people. One person wants to lose weight. The other wants to gain weight. In your dialogue, explain how they can help each other plan to have a healthful weight.

Your Healthful Weight

How do you know your healthful weight? The number you get from a scale is only part of the story. Your height and age and the shape of your body also affect your healthful weight.

Body Mass Index Your *Body Mass Index*, or BMI, is a number found by using a formula to compare your weight to your height. A doctor can use your BMI to help find your healthful weight.

Myth Type 2 diabetes is an adult disease.

Fact Type 2 diabetes is on the rise in children. It is related to a rise in obesity in children. Follow a daily exercise plan and a healthful diet to reduce the risk of Type 2 diabetes throughout your life.

► **Balance the calories you eat with the calories you use to maintain a healthful weight.**

Cade Martin/Getty Images

How to Gain and Lose Weight

Have a medical checkup before you decide to gain or lose weight. A doctor can find your BMI and take other measurements to determine your healthful weight. A doctor or nurse can use a special tool called calipers (CAL•i•perz) to measure the amount of fat under your skin. If you and your doctor agree that you need to lose or gain weight, your doctor can help you make a plan.

To lose weight, use more calories than you take in. You can do this by eating fewer calories and getting lots of physical activity. Physical activity burns calories and also helps you feel good about yourself.

To gain weight, take in more calories than you use. You can do this by eating more calories. You also can choose physical activities that help you build muscle. Having more muscle adds weight to your body.

ACTIVITY

Art LINK

Make a Collage

Make a collage showing ways to help maintain a healthful weight. From old newspapers and magazines, cut out pictures of healthful foods and different kinds of physical activity. Share your collage with your classmates.

What are some health risks associated with being overweight?

▲ **Exercises that build muscle can help a person gain weight.**

◀ **Exercises that burn calories can help a person lose weight.**

Body Image

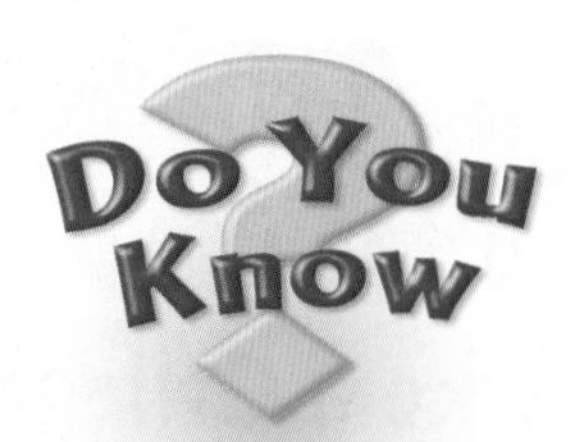

Myth Dietary substitutes, like drinks containing nutrients, are healthful.

Fact There is no substitute for eating a balanced diet. Using dietary substitutes can interfere with growth and development. A growing body needs a daily supply of nutrients. You must have approval from parents and a physician before using dietary substitutes.

Your **body image** is the feeling you have about the way your body looks. When you like and accept the way you look, you have a positive body image. Here are some ways to keep a positive body image.

Analyze Messages in the Media

Messages in newspapers, magazines, television, and movies can affect your body image. The pictures you see may make you think that one kind of body is better than another. To keep a positive body image, try not to compare yourself to other people.

Avoid Fad Diets

A *fad* is something that is popular for a short amount of time. Some weight-loss plans are called fad diets. Many fad diets promise that people will lose weight fast. Some do not include all the nutrients you need. A plan that combines healthful foods and physical activity is the best way to maintain a healthful weight.

Make Responsible Decisions

A friend tells you that she has decided not to eat, so that she can be thinner. She asks you not to tell anyone. What do you do?

1. **Identify your choices.** You could keep your friend's secret. What other choices do you have?
2. **Evaluate each choice. Use the *Guidelines for Making Responsible Decisions™*.** Ask yourself each question.
3. **Identify the responsible decision. Check this out with your parent or trusted adult.** Which decision leads to a "yes" answer to each question?
4. **Evaluate your decision.** Write a paragraph explaining your decision and why it is responsible.

Guidelines for Making Responsible Decisions™

- **Is it healthful?**
- **Is it safe?**
- **Does it follow rules and laws?**
- **Does it show respect for myself and others?**
- **Does it follow family guidelines?**
- **Does it show good character?**

Understand Eating Disorders

An **eating disorder** is a harmful way of eating because a person cannot cope with a situation. People who have eating disorders may have trouble dealing with problems or expressing feelings. They use food to try to feel better.

Binge Eating Disorder People who have binge eating disorder eat too much food all at once. They may overeat when they are hungry, angry, lonely, sad, or tired.

Bulimia Nervosa People who have bulimia (boo•LEE•mee•uh) eat a lot of food and then try to get rid of the food. People who have bulimia might throw up, or take drugs that cause bowel movements.

Anorexia Nervosa People who have anorexia (an•uh•REK•see•uh) eat very little or nothing at all. They usually think they are overweight, even when they are very thin.

Eating disorders are serious medical problems. People who binge are often overweight. People who have anorexia and bulimia do not get the nutrients they need. Their bodies may be malnourished.

People who have eating disorders need a doctor's help. They may need to go to the hospital. Counseling may help them develop a positive body image.

▲ **Do you have a positive body image?**

What is bulimia?

LESSON REVIEW

Review Concepts

1. **Explain** how to maintain a healthful weight.
2. **Explain** how to find your healthful weight.
3. **Describe** the causes, signs, and treatment of eating disorders.

Critical Thinking

4. **Infer** A TV show has seven characters. Five are very thin and seem popular. Two are overweight and not as popular. What messages might this show send?
5. LIFE SKILLS **Make Responsible Decisions** A friend has lost weight using a fad diet. He says that you should try it, too. What should you do?

CHAPTER 4 REVIEW

Use Vocabulary

- **balanced diet,** *B38*
- **body image,** *B60*
- **calorie,** *B40*
- **eating disorder,** *B61*
- **MyPlate,** *B38*
- **food group,** *B38*
- **Nutrition Facts label,** *B46*
- **pathogen,** *B51*

Choose the correct term from the list to complete each sentence.

1. Foods that contain similar nutrients make up a(n) __?__.
2. A harmful way of eating because a person cannot cope is a(n) __?__.
3. The unit used to measure energy in food is the __?__.
4. A germ that causes diseases is a(n) __?__.
5. You are told how much of each kind of food you need in the __?__.
6. Information about the nutrients in a packaged food is on the __?__.
7. The feeling you have about the way your body looks is your __?__.
8. A daily eating plan that includes the correct number of servings from the food groups is a __?__.

Review Concepts

Answer each question in complete sentences.

9. List three ways to limit fat in your diet. How does limiting fat help keep you healthy?

10. List the six kinds of nutrients your body needs. Explain why your body needs each kind of nutrient.

11. Name five foods from one of the food groups in MyPlate.

12. What foods are necessary to eat but are not included in MyPlate? Explain why you do not need large amounts of them in your diet.

13. What are combination foods? Give two examples.

Reading Comprehension

Answer each question in complete sentences.

There are two parts to weight management. You need to get physical activity. You also need to eat the correct amounts of healthful foods. Food labels usually say how many calories one serving of the food contains. Food labels also tell you how much fat, salt, and sugar are in the food. You can use this information to plan healthful meals.

14. What are the two parts of weight management?

15. List three kinds of information you might find on the Nutrition Facts panel on a food label.

16. How can reading food labels help you manage your weight?

Critical Thinking/Problem Solving

Answer each question in complete sentences.

Analyze Concepts

17. How might an ad in a magazine influence your body image?

18. Explain how you can use MyPlate to plan a balanced diet.

19. You are making dinner for your family. You are chopping vegetables for salad. You also need to cut up raw chicken. There is only one cutting board. What steps should you take to prepare the meal safely?

20. You and your doctor agree that you need to gain weight. How can you do it in a healthful way?

Practice Life Skills

22. **Analyze What Influences Your Health** You are watching television. You see many food ads that include professional athletes. Why might a company use athletes in its ads? What influence might this ad have on you?

23. **Make Responsible Decisions** The ice-cream truck is on your block. The other kids are running out to meet it. You want to go, too, but your parents or guardian promised to take you out for ice cream after dinner. What should you do? Use the Guidelines for Making Responsible Decisions™ to help you decide.

Read Graphics

The table shows the foods that Tonisha, Gabe, and Matt ate for one day.

Use it to answer questions 23–26.

	Tonisha	Gabe	Matt
Breakfast	Bagel with cream cheese	Toast with butter and jam	Nothing
Lunch	Tuna sandwich Apple juice 2 cookies	Bologna sandwich Yogurt	Fries Soda Candy bar
Dinner	Bean and cheese burrito Salad 1 slice of melon Milk	Spaghetti Garlic bread Soda	Fried chicken Onion rings Coleslaw Chocolate ice cream
Snacks	Frozen yogurt Candy bar	Cake Pretzels	Apple Orange

23. Who had the most balanced breakfast? Why?

24. How might Matt improve his diet to make it more heathful?

25. What might Gabe add to his diet to make it more healthful?

26. Who ate the most healthful snacks? Why were those snacks more healthful than the others?

UNIT B ACTIVITIES AND PROJECTS

Effective Communication

Write a Commercial

Write a radio script to advertise your favorite healthful food. Think up a jingle or a slogan. Include information about how it helps your body systems. Present your commercial to your class.

Self-Directed Learning

Write a Report

Choose one body system you have studied. Use the library to find information about the system. Write and illustrate a report about what you learn.

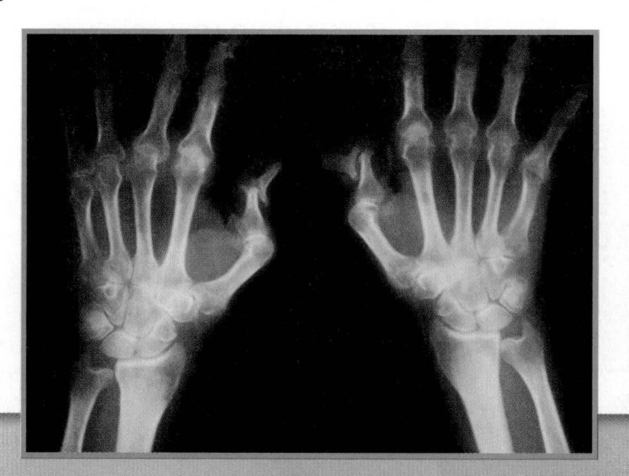

Critical Thinking and Problem Solving

Make a Chart

Down the left side of the chart, list the six major nutrients. Across the top, list the body systems. In each space, write how the nutrient affects that body system.

Responsible Citizenship

Take a Trip

With a responsible adult, visit a group that provides food to needy families. Speak with the staff about the types of food they provide. Offer to help with a chore.

UNIT C

Personal Health and Safety

CHAPTER 5

Personal Health and Physical Activity, *C2*

CHAPTER 6

Violence and Injury Prevention, *C40*

CHAPTER 5

Personal Health and Physical Activity

Lesson 1 • Caring for Your Body *C4*

Lesson 2 • Your Teeth, Eyes, and Ears *C10*

Lesson 3 • The Benefits of Physical Activity *C16*

Lesson 4 • A Balanced Workout *C22*

Lesson 5 • Play It Safe *C30*

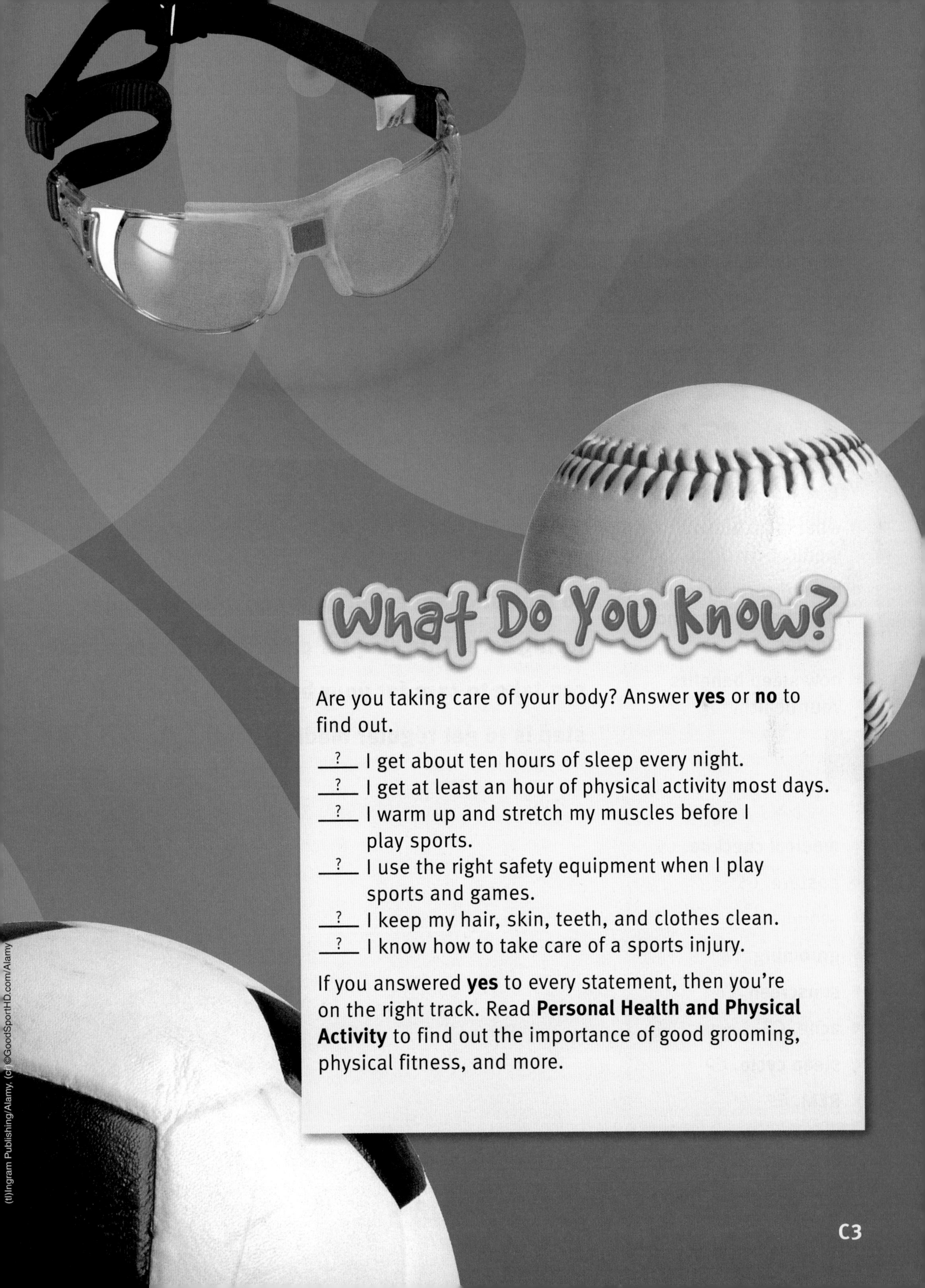

What Do You Know?

Are you taking care of your body? Answer **yes** or **no** to find out.

__?__ I get about ten hours of sleep every night.
__?__ I get at least an hour of physical activity most days.
__?__ I warm up and stretch my muscles before I play sports.
__?__ I use the right safety equipment when I play sports and games.
__?__ I keep my hair, skin, teeth, and clothes clean.
__?__ I know how to take care of a sports injury.

If you answered **yes** to every statement, then you're on the right track. Read **Personal Health and Physical Activity** to find out the importance of good grooming, physical fitness, and more.

LESSON 1

Caring for Your Body

You will learn . . .

- what is included in a medical checkup.
- how to care for your skin, hair, nails, and clothes.
- how sleep benefits your health.

Vocabulary

- **medical checkup**, *C5*
- **posture**, *C5*
- **scoliosis**, *C5*
- **grooming**, *C6*
- **sunscreen**, *C6*
- **acne**, *C7*
- **sleep cycle**, *C9*
- **REM**, *C9*

To look and feel good, your body needs tender loving care. This chapter describes the steps you can take to care for your body. One important step is to get regular medical checkups.

Medical Checkups

You should have regular medical checkups by your doctor. A **medical checkup** is a series of tests that measure your health status. The doctor checks that you are growing normally. The doctor checks for health problems. The earlier a problem is found and treated, the greater the chances of getting better.

You should have a medical checkup at least once a year. During the checkup, the doctor checks your heart, lungs, eyes, and ears to make sure that they are in good condition. The doctor or nurse measures your height and weight. Then the doctor tells your parents or guardian the results of the checkup.

Posture

During a checkup, your doctor or nurse may look at your posture. Your **posture** is the way you hold your body as you sit, stand, and move. You use muscles in your back and abdomen to sit and stand up straight. The *abdomen* (AB•duh•muhn) is the part of the body between the chest and the hips.

Correct posture helps your muscular and skeletal systems. If you slouch or hunch over, your spine and your muscles aren't in their proper places. Incorrect posture may be a sign of scoliosis. **Scoliosis** (skoh•lee•OH•sis) is a curving of the spine to one side of the body.

Why is a medical checkup important?

CRITICAL THINKING

Access Health Facts

Help your parents or guardian keep a personal health record for you.

1. **Identify when you might need health facts.** Keeping a record of facts about your health can help you, your parents or guardian, and a doctor make sure you grow and develop well.
2. **Identify where you might find health facts.** Talk to your parents or guardian about your health history.
3. **Find the health facts you need.** In a file or folder, include information about your most recent checkup, any medicines you take, and any allergies you have. Include your height and weight. Date all the information.
4. **Evaluate the health facts.** The next time you have a checkup, ask the doctor to read your record and tell you what else to include.

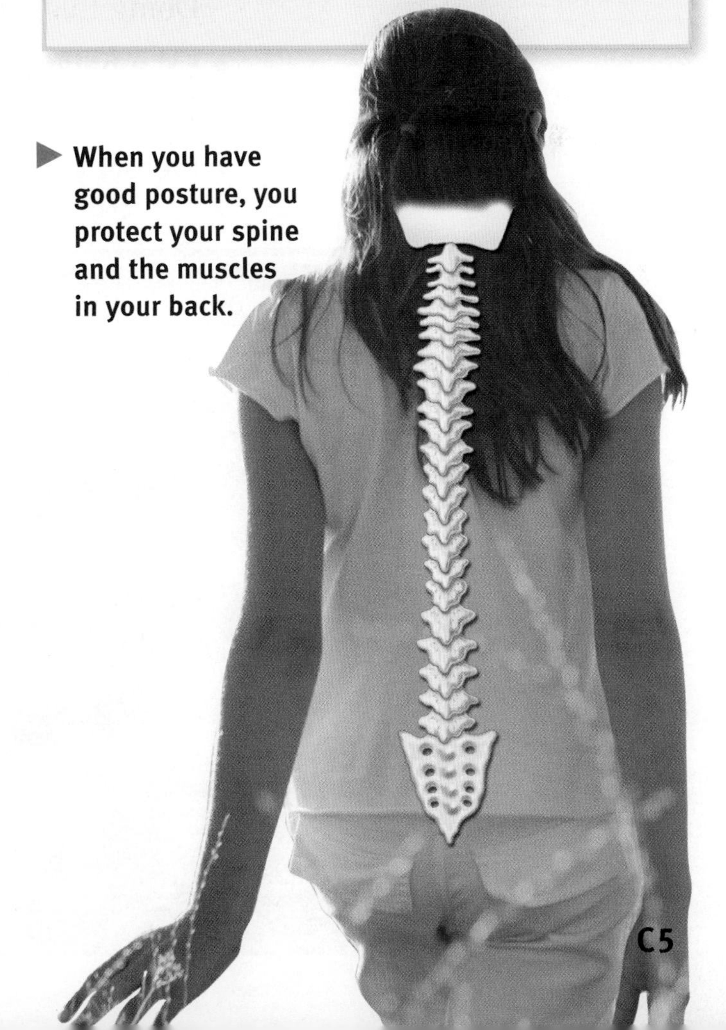

▶ **When you have good posture, you protect your spine and the muscles in your back.**

S. Olsson/PhotoAlto

See What Skin Does

How is your skin like an apple peel? Use a plastic knife to peel an apple. Leave the peel on another apple. Set both apples on a tray and leave them for half an hour. What happens to the peeled apple? Compare it with the intact apple. Your skin protects your body, just as the peel protects an apple from turning brown in the air.

Good Grooming

Grooming is taking care of your body and appearance. Being well-groomed is having a neat, clean appearance. Being well-groomed can help you feel good about yourself and the way you look.

Skin Care

Your skin is your largest organ. It covers your body and protects it from germs. Your skin helps control your body temperature. It keeps moisture inside so that your internal organs do not dry out. When your body temperature rises, you sweat through your skin. This helps cool your body.

Keeping your skin clean and healthy is part of good grooming. Here are some ways you can keep your skin healthy.

- **Wash your skin** with soap and water at least once a day. Doing this removes dirt, oil, germs, and sweat.
- **Eat a balanced diet** with lots of fruits and vegetables. This gives your body the nutrients your skin needs.
- **Protect your skin** from the sun. The sun gives off rays called ultraviolet A (UVA) and ultraviolet B (UVB) that can harm your skin. These rays also increase your risk of skin cancer. **Sunscreen** is a lotion or cream that blocks the sun's harmful rays. Use a sunscreen with a Sun Protection Factor (SPF) of at least 15 every time you go outside. Make sure it protects against both UVA and UVB rays. Use at least an ounce of sunscreen each time you apply it. Wear a hat and protective clothing, too.

▶ **Skin care and hair care products help you stay well groomed. Always read the label of a product to make sure you are using it properly.**

Your Skin and Puberty

During puberty your skin produces more oil. The oil can trap germs and dead cells in *pores,* or tiny openings in the surface of your skin. Clogged pores can cause pimples and other skin problems. **Acne** is a skin disorder in which clogged pores become inflamed, or swollen. Washing your face every day can help keep the pores in your skin clean and open.

During puberty, you perspire more, especially under your arms. Bacteria in perspiration can cause odor. Bathing at least once a day can help control odor. Grooming products called *deodorants* help control odor. *Antiperspirants* reduce the amount of perspiration your skin produces.

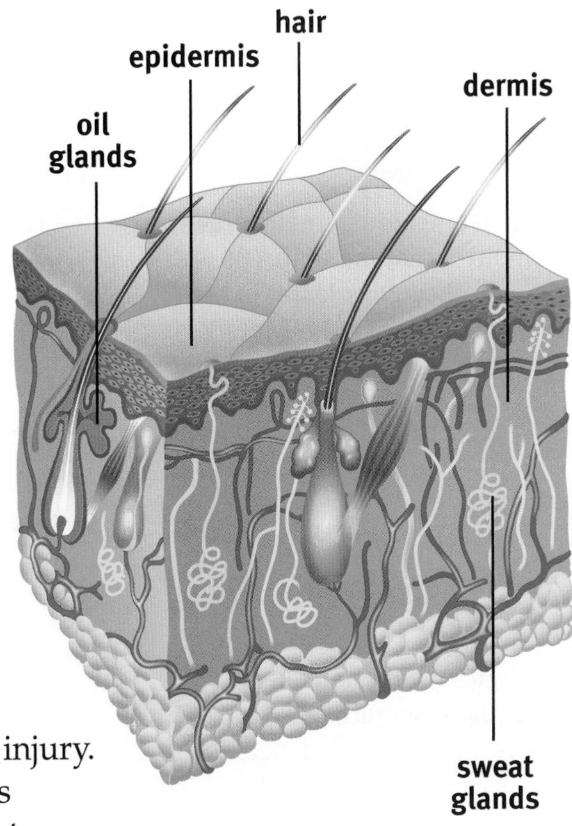

▲ **Your skin has two main layers, the dermis and epidermis. Oil and sweat are produced in glands in the dermis.**

Hair Care

Your hair covers your scalp and protects it from injury. It also protects the skin on your head from the sun's harmful rays. Washing your hair regularly helps it stay clean and healthy. You may want to use conditioner if your hair is hard to comb. Conditioner helps remove tangles from hair. Comb or brush your hair every day to help it stay tangle-free.

Even clean, healthy hair can have problems. *Dandruff* is a condition in which pieces of dry or greasy skin from your scalp flake off into your hair. Medicated shampoos can help reduce dandruff.

Head lice are tiny insects that lay eggs in hair. The lice make your scalp itch and cause sores to appear. You can help reduce your risk of getting head lice. Don't share hats, brushes, or combs with others. Head lice can pass from one person to another this way. If you do get head lice, special shampoos can kill the lice and their eggs.

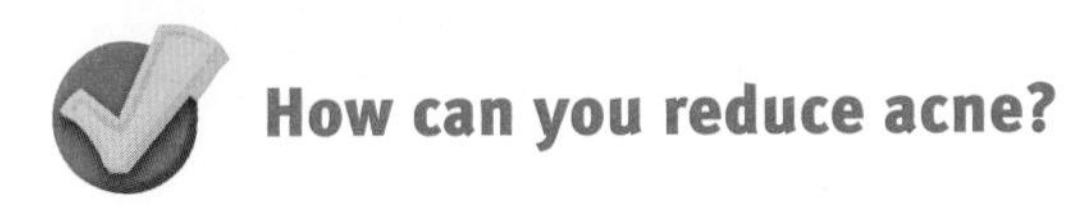

How can you reduce acne?

On average, fingernails grow about 1 centimeter in 100 days. They grow faster in the summer than in the winter!

Nail Care

Good grooming includes keeping your fingernails and toenails clean and trimmed. When you wash your hands, remember to clean under your fingernails. You can use a nail brush to scrub your nails gently.

Trim and file your nails as needed. This keeps them from snagging on clothes. It also makes them less likely to break. Trim your nails straight across. Do not clip them too close to your fingertips. Use a nail file to smooth out any rough edges.

Biting your nails increases the risk of illness because you put your hands in or near your mouth. This allows germs into your mouth. Your nail beds, the area under the nails, can get infected if the skin around them is torn. The torn skin lets germs get under the nails. Tell your parent or guardian if the skin around your nails swells, bleeds, or hurts. This can be a sign of infection.

▼ **Filing your nails keeps them smooth.**

Care for Your Clothes

You look your best when your clothes are clean. Clean clothes also smell better than dirty ones. Make sure that your clothes are washed regularly. You may help your parents or guardian do the laundry. Wash your clothes using laundry detergent. Make sure that your clothes are completely dry before you put them away.

You should wash your sheets and towels often, too. Dirty sheets and towels can spread diseases and head lice.

Why is it important to wash your sheets regularly?

Ken Karp/McGraw-Hill Education

Rest and Sleep

Getting plenty of sleep and rest is an important part of staying healthy. As you sleep your body repairs cells and grows. When you are well rested, your body is better able to resist disease. You can concentrate better in school.

Babies need at least 16 hours of sleep per day. You need about 10 hours of sleep. Most adults need 7 or 8 hours of sleep.

The stages your body goes through during sleep are called the **sleep cycle**. The sleep cycle consists of the five stages shown below. The longest stage of the sleep cycle is rapid eye movement sleep, or REM. **REM** is the stage of sleep when you dream. During this stage your eyes move very quickly, but your body does not move at all.

What is the sleep cycle?

▼ **It takes about an hour and a half to two hours to pass through all five stages of sleep.**

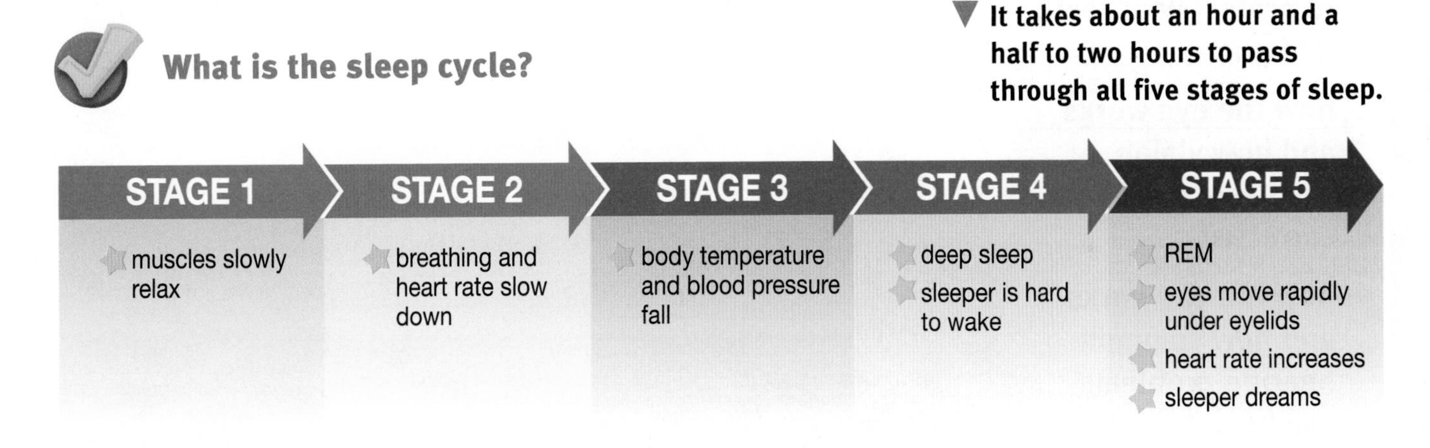

LESSON REVIEW

Review Concepts

1. **Identify** what is included in a medical checkup.
2. **Describe** how to care for your skin, hair, nails, and clothes.
3. **List** three benefits of getting enough sleep.

Critical Thinking

4. **Summarize** How does good grooming promote health?
5. LIFE SKILLS **Access Health Facts** Your friend's doctor told your friend that she has scoliosis. Where could she find more information about this condition?

LESSON 2

Your Teeth, Eyes, and Ears

You will learn . . .

- how the teeth function and ways to protect your teeth and gums.
- how the eye works and how vision problems can be corrected.
- how the ear works and how common hearing problems can be detected and avoided.

Vocabulary

- **periodontal disease**, *C11*
- **decibel**, *C15*

You use your teeth, eyes, and ears every day when dealing with the world around you. It's important to take care of them and keep them healthy.

Your Teeth

Teeth are made of a hard, strong tissue that contains calcium. The surface of your teeth is made of *enamel,* the hardest substance in your body. Enamel and a bonelike substance called *dentin* cover the *pulp,* the part of a tooth that holds nerves and blood vessels.

Dental plaque is a sticky substance on teeth that contains bacteria. Plaque combines with sugar to make an acid. The acid can make holes called *cavities* in your teeth.

To protect your teeth, avoid sugary foods. Brush and floss your teeth every day. Flossing removes dental plaque between teeth. Brushing and flossing also help prevent periodontal disease. **Periodontal** (per•ee•uh•DAHN•tuhl) **disease** is a disease of the gums and bone that support the teeth. It can cause tooth loss.

Regular dental checkups are important. A dentist or dental hygienist can clean away any plaque buildup. The dentist can fill any cavities. The dentist also checks that your teeth are straight. The dentist might suggest that you see an *orthodontist,* a dentist who is trained to fit braces on teeth. The dentist also can give you a mouth guard to protect your teeth when you play sports. You should have a dental checkup every six months.

What is periodontal disease?

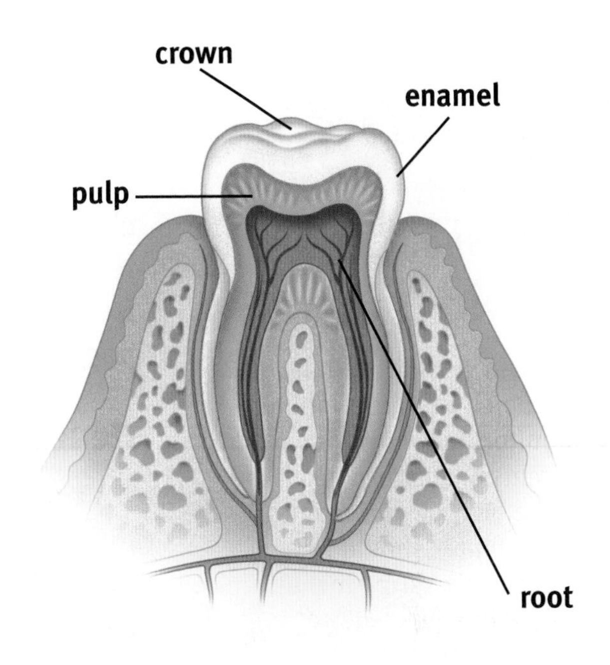

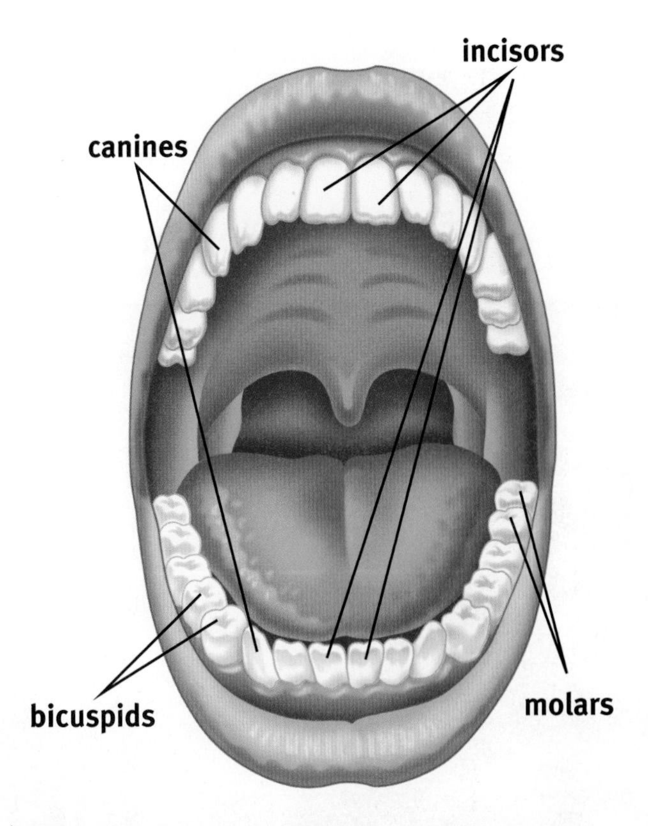

▲ **You have four kinds of teeth. Incisors (in•SIGH•zurz) cut food. Canines (KAY•nighnz) cut and tear food. Bicuspids (bi•KUS•pidz) crush and grind food. Molars (MOH•lurz) grind food before you swallow it.**

Your Eyes

ACTIVITY

On Your Own

FOR SCHOOL OR HOME

Avoid Eyestrain

You can get eyestrain if you sit too close to a computer monitor. Talk with your parents or guardian about the placement of your monitor. The monitor should be placed at least 24 inches from where you sit.

Your eyes bring you information about the outside world. Light enters your eye through your *pupil,* a hole in the center of your eye. It passes through the *lens,* a curved, clear part of the eye that focuses images. Then it hits the retina. The *retina* is the back of your eye. It sends messages from your eye to your brain. The diagrams on the next page show the parts of the eye.

Healthy Eyes

You can form good habits to keep your eyes healthy. First, be sure to have a doctor check your vision each year. Your eye doctor can find and treat any problems early. If your eyes ever itch or burn, tell your parent or guardian right away. Your eyes might be infected.

Your eyes get tired, or strained, if they work too hard. To prevent this, read and watch television in well-lit areas. Rest your eyes every few minutes when you use a computer. Look at something far away to give your eyes a break. Don't rub your eyes, even when they're tired. You can spread an infection or damage your eyes.

The sun can damage your eyes. Never look directly at the sun. Wear sunglasses that protect you from the sun's UVA and UVB rays.

You can reduce your risk of eye injuries. Wear safety glasses when you play sports that require them. Never run with or throw anything sharp or pointed. Carry sharp objects with their points down.

Sitting at the right distance from a computer can protect your vision.

Vision Problems

Some people have trouble seeing things nearby or far away. This happens because their eyes have an irregular shape. The diagrams below show how the shape of the eyeball affects vision. Eyeglasses or contact lenses can help correct vision. *Contact lenses* are lenses that fit directly on the eye. An eye doctor can give you a prescription for glasses or contact lenses if you need them.

Laser surgery is another way to correct vision. Doctors use a powerful beam of light to reshape the eyeball.

What should you do if your eyes itch or burn?

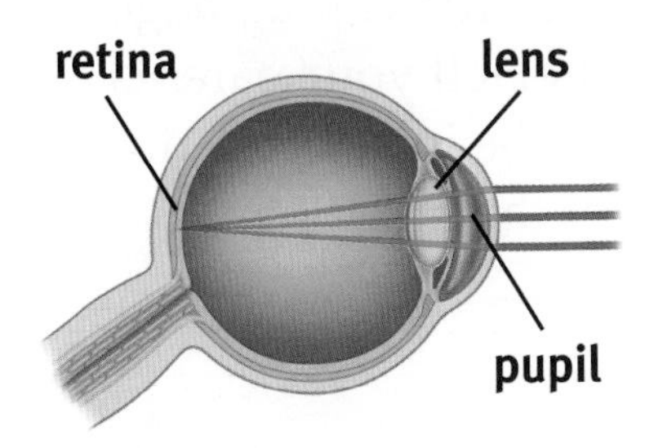

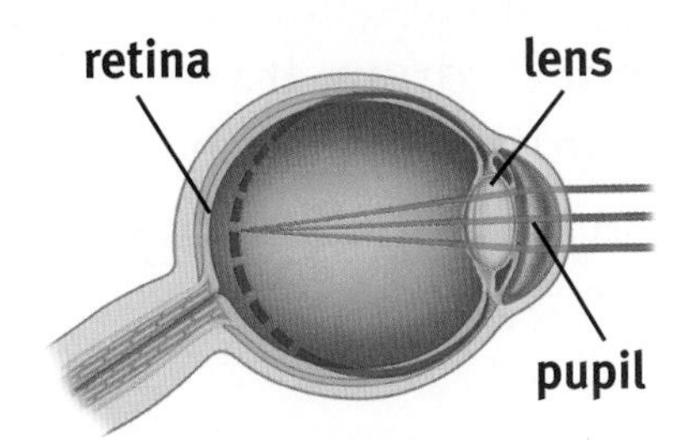

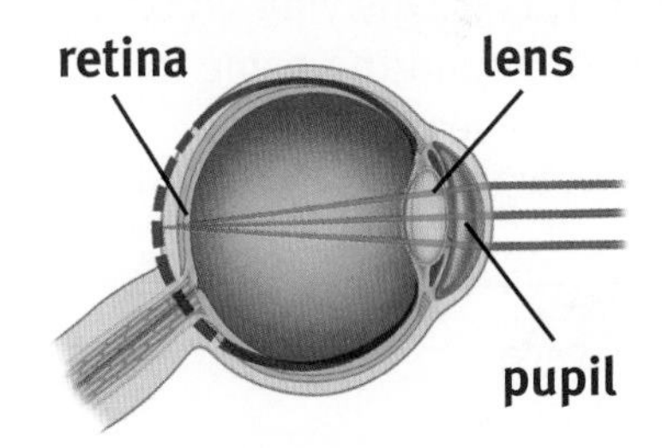

▲ In a person with normal vision, the lens focuses images clearly on the retina.

▲ A person who is *nearsighted* sees nearby objects clearly. Faraway objects look blurry. The image is focused in front of the retina.

▲ A person who is *farsighted* sees faraway objects clearly. Nearby objects look blurry. The image is focused behind the retina.

Your Ears

ACTIVITY

Science LINK

Tape Your Voice

Use a tape recorder to record yourself saying a few sentences. Play it back. Does it sound like your voice to you? When you hear your own voice, you hear it through the bones, muscles, and other tissues in your head. This is why your voice may not sound the same to you as it does to others.

What happens when you listen to a song on the radio? The sound of the music enters your ear through the *outer ear.* The part of your outer ear that you can see helps funnel the sound into your ear canal. Your *ear canal* carries the sound to your *eardrum,* a thin membrane that vibrates when hit by sound. The vibrations move through three small bones to the cochlea (KOH•klee•uh). The *cochlea* is a shell-shaped part of your ear that is filled with fluid. When the fluid vibrates, it causes nerves to send messages to your brain about what you are hearing. Your brain interprets the messages.

It's important to protect your ears. Never put anything in your ears, not even a cotton swab. You can damage your eardrum. If you get an earache, tell your parents or guardian right away. Your ear could be infected. A doctor can give you medicine to get rid of the infection.

ACTIVITY

LIFE SKILLS

CRITICAL THINKING

Make Responsible Decisions

Your brother asks you to turn up the volume on the stereo because he likes loud music. You want to protect your hearing. What should you do?

1. **Identify your choices.** You could turn up the volume. You could keep the volume low.
2. **Evaluate each choice. Use the *Guidelines for Making Responsible Decisions*™.** Ask yourself each question for each of your choices.
3. **Identify the responsible decision. Check this out with your parent or trusted adult.** Which decision protects your hearing?
4. **Evaluate your decision.** Write a paragraph explaining your decision and why it is responsible.

Guidelines for Making Responsible Decisions™

- **Is it healthful?**
- **Is it safe?**
- **Does it follow rules and laws?**
- **Does it show respect for myself and others?**
- **Does it follow family guidelines?**
- **Does it show good character?**

Hearing Loss

Your doctor may test your hearing when you have a checkup. A hearing specialist called an *audiologist* (aw•di•AH•luh•jist) may do the test. The test checks to see whether you have hearing loss. *Hearing loss* is the inability to hear or interpret certain sounds. People who have hearing loss may use hearing aids. Some have surgery to help them hear better. Others learn sign language.

Have you ever heard a noise that was so loud that it made your ears ring? Repeated exposure to loud sounds can cause hearing loss. The measure of the loudness of a sound is a **decibel**. A normal conversation has a loudness of about 60 decibels. Sounds louder than 85 decibels, such as a jet engine, a lawnmower, or even loud music, can damage your hearing permanently.

The more time you spend around loud sounds, the more your hearing can be damaged. Keep music at a low volume. Protect your hearing when you must be around loud sounds. Wear earplugs to reduce sound levels. When you listen to music with headphones, keep the volume low. If anyone else can hear the sound, it is too loud.

▲ **Ear protectors can help protect your ears from loud sounds. These protective devices can reduce the loudness of a sound by 15 to 30 decibels.**

What is a decibel?

LESSON REVIEW

Review Concepts

1. **Describe** three ways to keep your teeth healthy.
2. **Explain** the structure and function of the eye.
3. **Explain** the structure and function of the ear.

Critical Thinking

4. **Contrast** What is the difference between being nearsighted and being farsighted? How can each be corrected?
5. LIFE SKILLS **Make Responsible Decisions** You and a friend are planning to go to a rock concert. The music will be very loud. You want to bring earplugs to reduce the volume. Your friend thinks they will look stupid. What should you do?

LESSON 3

The Benefits of Physical Activity

You will learn . . .

- about the social, emotional, and physical benefits of physical activity.
- about the five kinds of health fitness.
- about the six kinds of fitness skills.

Vocabulary

- **blood pressure**, *C17*
- **health fitness**, *C18*
- **heart rate**, *C18*
- **cardiac output**, *C18*
- **fitness skills**, *C20*
- **President's Challenge**, *C20*

Do you enjoy outdoor activities? How do you feel when you play hard? Physical activity can make you feel great. It keeps your body healthy, too.

Physical Fitness

Do you walk, jog, or swim? Do you play sports? Physical activities help you stay fit. *Physical fitness* means having your body in top condition.

Physical activity keeps your body strong. It gives you energy, too. It reduces the risk of disease later in life. Being fit helps you maintain a healthful weight. It also reduces the risk of high blood pressure. **Blood pressure** is the force of blood against artery walls. High blood pressure harms your arteries.

Physical activity helps your mental and emotional health, too. It helps reduce the harmful effects of stress. It helps you feel better about yourself. This improves your self-concept.

Physical activity can benefit your family and social health. You can make new friends while you play sports. You can enjoy physical activities with your family.

How does physical fitness benefit your emotional health?

Write About It!

Write a Story Write a story about a sport or other physical activity you've learned and worked at. Was it challenging? How did your body adapt over time? Did it become easier as you became more physically fit?

▶ Some popular physical activities come from other cultures, such as this East Asian activity. Trying a new physical activity may be hard at first. If you keep at it, though, your body will become stronger. Find a friend to practice with. What physical activity would you like to try?

Health Fitness

Health fitness is having your heart, lungs, muscles, and joints in top condition. There are five kinds of health fitness.

▲ **Running, swimming, and playing basketball can all increase your cardiorespiratory endurance.**

- **Cardiorespiratory endurance** (KAR•dee•oh•REHS•per•uh•taw•ree) is the ability to stay active without getting tired. When you run, swim, or dance, you probably feel your heart rate increase. Your **heart rate** is the number of times your heart beats each minute.

  Your heart works hard to get oxygen to your muscles. Your **cardiac output** is the amount of blood pumped by your heart each minute. It increases when you have good cardiorespiratory endurance.

  You can increase your cardiorespiratory endurance with aerobic activities. *Aerobic* (ayr•OH•bik) *activities* make your heart beat faster but at a steady rate. Running, dancing, and playing soccer are all aerobic activities.

- **Muscular endurance** is the ability to use the same muscles for a long time. You can help build muscular endurance by carrying a grocery bag or your textbooks, or by doing pull-ups.

- **Flexibility** is the ability to bend and move your body easily. Gymnastics, stretching exercises, and ballet can help you increase your flexibility.

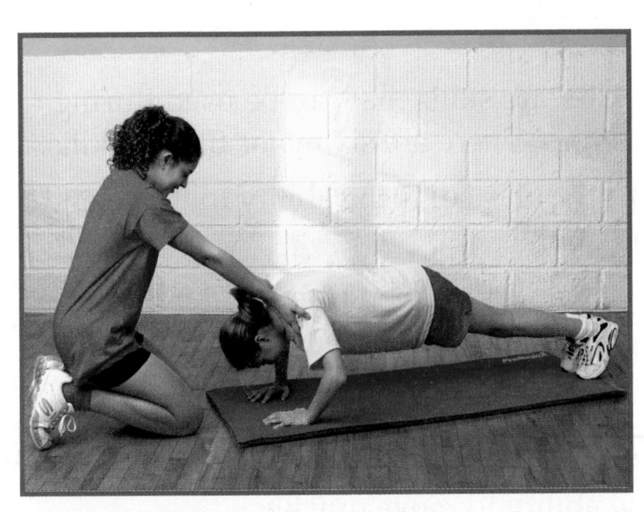

▲ You can improve your muscular strength by doing push-ups.

Do You Know

In March 2003, a man from Florida set a world record by doing 133 push-ups in just one minute. How many can you do?

- **Body composition** is the amount of fat tissue and lean tissue in your body. Your body uses fat to protect your organs from injury. Fat insulates your body to keep you from losing heat. Fat is also used as energy when your body needs it. Lean tissue includes muscles, bones, nerves, skin, and organs. To be fit you need to have a proper balance of fat tissue and lean tissue in your body. When you are fit, you have less fat tissue and more lean tissue.
- **Muscular strength** is the amount of force your muscles can produce. Strong muscles help you lift, kick, push, and pull. Exercises such as push-ups and pull-ups can help increase your muscular strength.

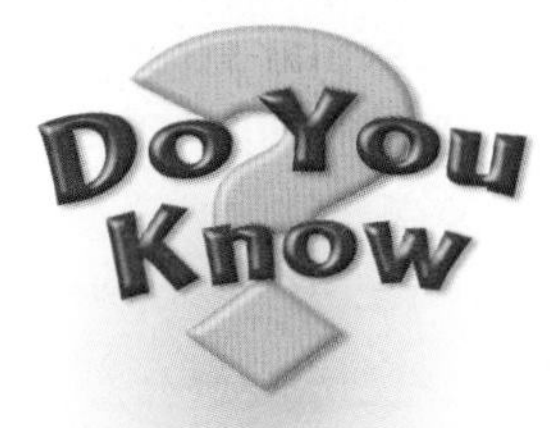

Myth All sports drinks benefit the body.

Fact Many sports drinks contain sugar and artificial ingredients, substances which are not beneficial to health. Better stick with water.

What physical activities can you do to increase cardiorespiratory endurance?

◀ Flexibility is an important part of health fitness. Being flexible helps you do many activities.

Test Your Skills

Fitness skills are skills that can be used during physical activities. There are six major fitness skills.

- *Agility* (uh•JI•luh•tee) is the ability to move and change directions.
- *Balance* is the ability to keep from falling.
- *Coordination* (koh•or•di•NAY•shun) is the ability to use your body parts and senses together.
- *Reaction time* is the time it takes to move after a signal.
- *Speed* is the ability to move quickly.
- *Power* is the ability to combine strength and speed.

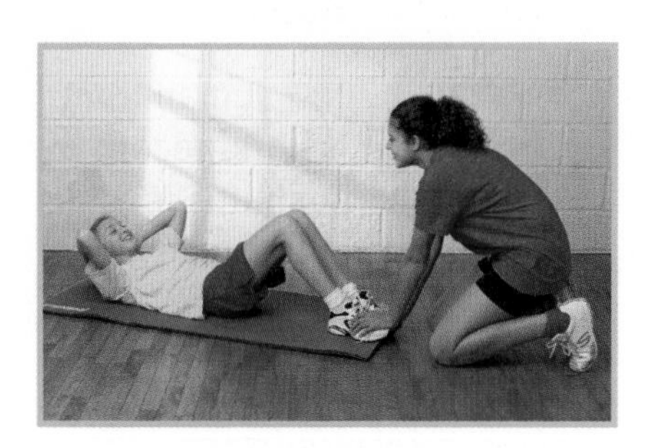

▲ Curl-ups test your muscular strength and endurance.

Test Yourself

There are tests that measure your physical fitness. One such test, the **President's Challenge**, is sponsored by the government. The President's Challenge includes a one-mile walk or run, curl-ups, pull-ups, V-sit and reach, and a shuttle run. There are three awards given to participants of the President's Challenge. Every person who participates in the challenge gets an award based on his or her results of the five tests.

▲ V-sit and reach tests your flexibility.

▲ One-mile walk/ run tests your cardiorespiratory endurance.

▲ Pull-ups test your muscular strength and endurance.

▲ Shuttle run tests your speed, reaction time, and agility.

Body Composition

Doctors often use body composition to determine if a person's weight is healthful. Body composition measures the proportion of a person's fat tissue to his or her lean tissue. Lean tissue includes muscles, bones, and organs.

To measure body composition, a doctor may use the "pinch test." This test is done by pinching different areas on one side of the body with special equipment called *calipers*. The doctor pinches the skin and takes a reading. The results are then plugged into a formula to determine your body composition.

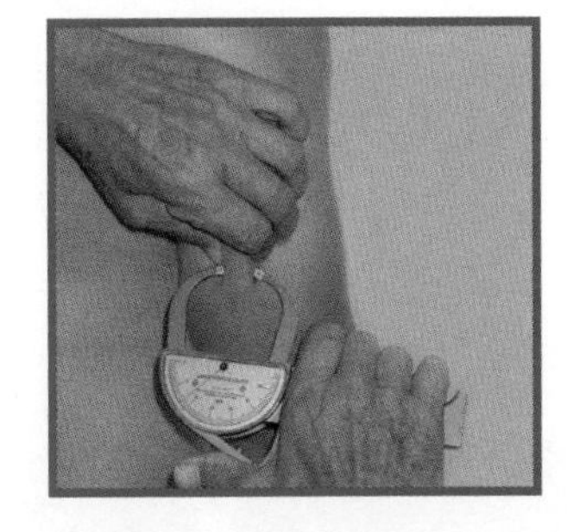

▶ **Body composition is tested with the "pinch test."**

What six fitness skills can be used during physical activity?

ACTIVITY
LIFE SKILLS
CRITICAL THINKING

Practice Healthful Behaviors

1. **Learn about a healthful behavior.** List the six fitness skills. With a small group, choose an activity you could do to improve each skill. The activities should be things you can do in your classroom or on the school playground. You might toss a ball between two people as fast as you can to practice coordination, for example.
2. **Practice the behavior.** Get permission from your teacher to try your activities. Use a chart to keep track of the results.
3. **Ask for help if you need it.** Ask your teacher for suggestions on how to improve your skills.
4. **Make the behavior a habit.** Practice your activities often. Do your fitness skills improve?

LESSON REVIEW

Review Concepts

1. **Explain** how physical activity benefits your health.
2. **List** the five kinds of health fitness.
3. **Identify** the six fitness skills.

Critical Thinking

4. **Synthesize** How can physical activity help your self-confidence and your body image?
5. LIFE SKILLS **Practice Healthful Behaviors** Make a plan to improve your fitness skills. Record your progress and evaluate the results.

LESSON 4

A Balanced Workout

You will learn . . .

- what amount and intensity of physical activity is healthful.
- kinds of physical activity to choose.
- how to work out safely and effectively.
- how to set health goals for fitness.

Vocabulary

- **target heart rate**, *C26*
- **specificity**, *C26*
- **training**, *C26*
- **overload**, *C26*
- **progression**, *C27*
- **frequency**, *C27*

Team sports are a great way to enjoy physical activity, but they're only one part of a balanced workout. You can plan an individual exercise routine to improve all areas of your health fitness.

Purestock/SuperStock

Benefits of Physical Activity

Suppose you play a game of tag with your friends. You may be playing just for fun, but you're also getting exercise. To *exercise* means to plan specific physical activities to reach a physical fitness goal.

Exercise strengthens your lungs and heart. It keeps your blood pressure at a healthful level. It strengthens your muscles and improves your flexibility. Being physically fit can benefit your mental and emotional health as well.

Intensity

Intensity describes how hard you work during physical activity.

- **Low-intensity physical activities** make your heart work a little harder than usual. Walking slowly and cleaning your room are low-intensity activities.
- **Moderate-intensity physical activities,** such as fast walking and swimming, make your heart and muscles work harder.
- **Vigorous-intensity physical activities,** include running, swimming laps, and bicycling over hills. These activities make your heart and muscles work even harder than moderate-intensity activities. Your heart rate, breathing rate, and blood pressure increase even more.

To be physically fit, spend at least 60 minutes in moderate-intensity activities each day. You also need 30 minutes of vigorous activity at least three or four times each week.

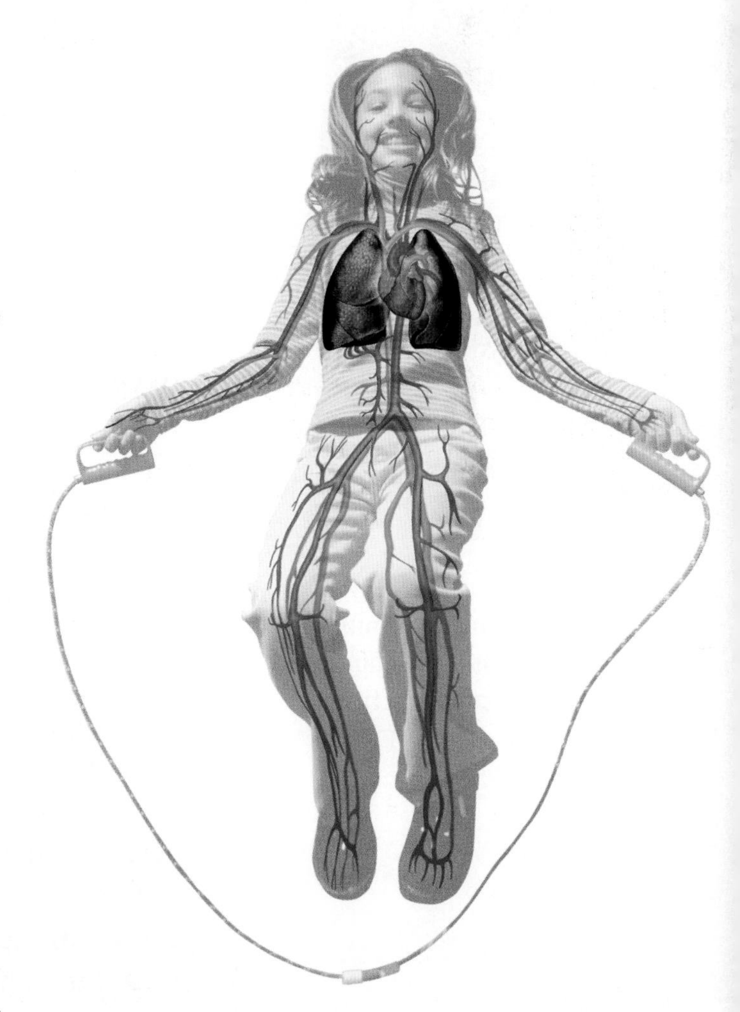

▲ **Exercise strengthens your heart and lungs.**

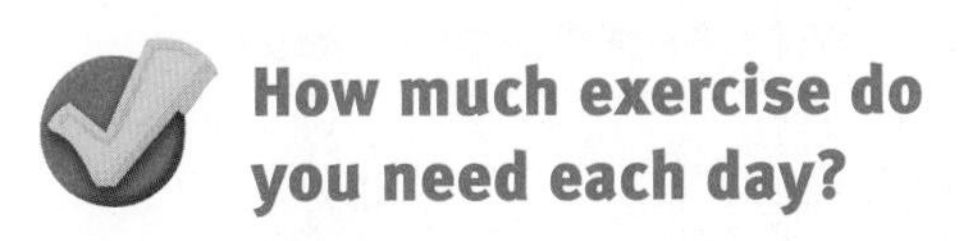

How much exercise do you need each day?

▲ **Improving your muscle strength is an important part of physical fitness.**

Getting Started

Check with your parents or guardian before you plan an exercise program. Get a medical checkup before you start, too. Talk to your doctor about the exercises you want to do. Ask questions. He or she can help you get started. A doctor can suggest activities suited to your age. There may be activities more suited to the needs of females or to the needs of males. A physical fitness plan should improve all five kinds of health fitness.

Kinds of Physical Activity

Aerobic exercise makes your body use large amounts of oxygen over a long time. Walking, jogging, bicycling, swimming, and in-line skating are all aerobic activities. Aerobic exercises improve your cardiorespiratory endurance.

Anaerobic (a•nuh•ROH•bik) exercises use short periods of hard work followed by periods of rest. During the exercise, your body needs more oxygen than it can get from your lungs and heart. Your heart beats faster to get oxygen to your muscles, but it can't keep up, so you stop to rest. Push-ups, pull-ups, sit-ups, and lifting weights are all anaerobic exercises.

Anaerobic exercises improve your muscle strength and endurance. The combination of aerobic and anaerobic exercises helps improve your body composition.

Stretching is another kind of exercise. When you stretch, you gently pull on your muscles. Stretching improves your flexibility. It also keeps your muscles from feeling tight after you do other exercises.

Sports games can be fun and exciting kinds of physical activity. In addition to baseball and football, you might look to sports games from other cultures. Soccer and cricket are just two examples of sports games that are popular in many countries around the world.

What is the difference between aerobic and anaerobic exercise?

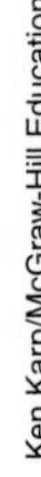

How to Choose

Choose physical activities and exercises carefully. Choose activities you like. If you like an activity, you are more likely to do it regularly.

No one exercise will improve all five kinds of health fitness and all six fitness skills. Choose a variety of exercises. For instance, suppose you decide to include swimming in your exercise plan. Swimming will improve your cardiorespiratory endurance, but it will not help your muscular strength as much. You could do push-ups and pull-ups to improve your muscular strength and endurance. You could also take a martial arts class to improve your flexibility, agility, and coordination. Find the right combination of exercises for you.

Check to see if the activities you choose have required equipment. Make sure you have safety equipment that works, fits you, and is safe.

Consumer Wise ACTIVITY

Make a Buying Guide

Suppose you need to buy running shoes. Write a list of things to consider before you buy them. The list may include questions such as these:

- How much money do I have to spend?
- How can I decide which style is best for me?

Review your list with a parent or guardian.

Your parents or guardian can help you choose safe sports equipment.

Work Out

Calculate Your Target Heart Rate

1. Subtract your age from 220.
2. Multiply the result by 0.6. This is the lower limit of your target heart rate.
3. Multiply the result of step 1 by 0.75. This is the higher limit of your target heart rate.
4. Now jog in place for 30 seconds. Find your heart rate using the instructions in the caption below. Does it fall in your target heart range? Before exercising, check with a doctor to see if you should use a different target heart rate.

Warm Up

A *warm-up* is three to five minutes of easy physical activity done before a workout. Walking and jogging slowly are good warm-ups.

Warm-ups raise your heart rate. They also increase the blood flow to your muscles. This helps your muscles get ready to work hard. Warming up also helps prevent injuries.

Work Out

After you warm up, you are ready to begin your work out. If you are doing aerobic exercise, check your heart rate. Your **target heart rate** is a fast and safe heart rate for workouts. It is fast enough to make your heart work harder and build your cardiorespiratory endurance.

You may know which areas of fitness you want to improve. **Specificity** is doing specific exercises and activities to improve particular areas of fitness. For example, if you want to improve the muscle strength in your arms, you can do push-ups and pull-ups. **Training** is doing specific exercises to improve a particular fitness skill or type of health fitness. For example, you could run sprints to improve your speed.

To improve your health fitness, you need to overload. In **overload** work your body harder than normal. As your body gets stronger, it takes more work to overload it. Your body gets used to working hard. You need to challenge it to work even harder. This will help your health fitness continue to improve.

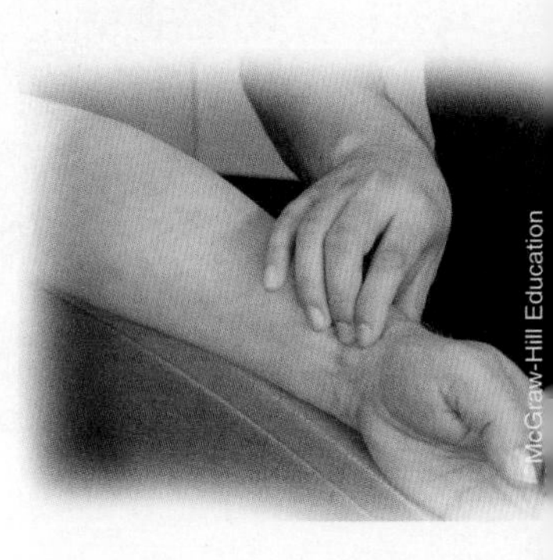

To find your heart rate while you work out, count the number of beats in 6 seconds and add a zero. This gives you the number of beats per minute. You can feel the beats by placing your index or middle finger on the inside of your wrist or the side of your neck.

Overload should not happen too quickly. This could make your muscles sore. You might risk injury. Instead, use progression. **Progression** is the gradual increase in overload necessary to achieve higher levels of fitness. Increase the frequency, intensity, or amount of time you spend doing an activity. **Frequency** is how often something happens. Working your body often helps increase your strength and endurance.

▲ **Drink plenty of water before, during, and after your workout.**

FITT

To remember how to work out effectively, remember the acronym FITT:
F requency Get physical activity every day.
I ntensity Exercise at a moderate to vigorous level of intensity.
T ime Get at least 60 minutes of physical activity every day.
T ype Choose the right types of exercise. For example, try stretching exercises, dancing, or gymnastics to improve flexibility.

Cool Down

Cool down when you finish exercising. A *cool-down* is five to ten minutes of easy physical activity done after a workout. You might walk or jog slowly. As you cool down, your heart rate and blood pressure return to normal.

Stretching at the very end of your cool-down helps you relax. Stretching also protects your muscles. It can improve your flexibility. Be sure to drink plenty of water after you exercise. As you perspire your body loses water. You need to replace that lost water as soon as possible.

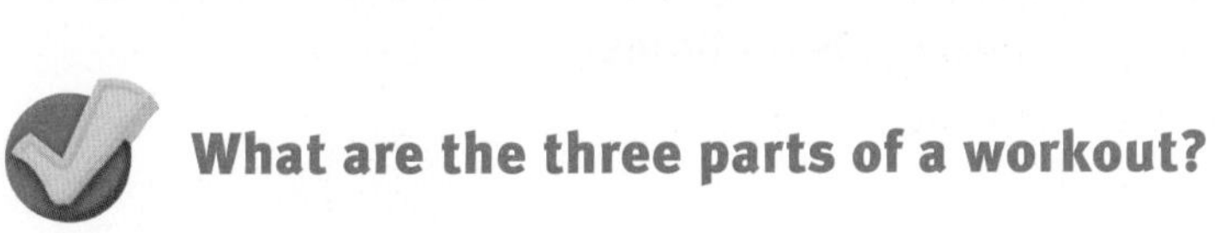

What are the three parts of a workout?

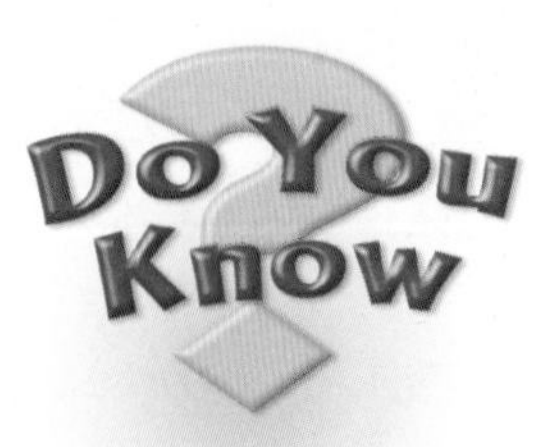

Myth Soft drinks do not cause health problems.

Fact Soft drinks can be thought of as liquid candy, because they contain substantial amounts of sugar. Sugar consumption has been linked with obesity in children.

Setting Health Goals for Fitness

Each person has different fitness needs. Your age, heredity, and level of fitness all affect your fitness needs. You should think about all these things when you set health goals for fitness. Your friends' health goals for fitness may be different from yours. Your culture may influence your choice. You may prefer to play soccer, for example, instead of American football. Choose the health goals that are right for you.

Your parents or guardian can help you set health goals for fitness. A doctor, nurse, or sports coach also can help you decide which physical activities will help you reach your health goals. These health professionals may suggest activities more suited to the physical needs of boys or of girls.

It takes time to be fit. You'll need frequent workouts to reach your goal. Make time for physical activity and exercise each day.

Sometimes you may get tired or frustrated. You may want to stop exercising. When this happens, remember your goal. Think about how much progress you have already made. Ask an adult or a friend for help and encouragement if you need it.

◀ **Making a schedule and managing your time wisely are the first steps toward meeting your health goals for fitness.**

Making Time

You probably have a busy life. You may have to balance school, after-school activities, homework, and time with your family. How can you include daily exercise, too?

You can use a computer to help you make a weekly schedule. You can keep track of your progress on a calendar, too. Tell who can help you—a parent or guardian, a family doctor, or a coach. Display your schedule in your room as a reminder.

Make room in your schedule for exercise by spending less time watching television and playing computer and video games. This will give you more time to do the things you need to do and still get in a good workout.

How can you track your progress toward meeting your health goals for fitness?

ACTIVITY

LIFE SKILLS

CRITICAL THINKING

Set Health Goals

1. **Write the health goal you want to set.** I will get plenty of physical activity. Write your goal in a Health Behavior Contract.
2. **Explain how your goal might affect your health.** Write a few sentences telling how getting physical activity will improve your health.
3. **Describe a plan you will follow to reach your goal. Keep track of your progress.** Design a plan to improve one area of health fitness. Write your plan into your Health Behavior Contract.
4. **Evaluate how your plan worked.** Did the physical activities improve the area of health fitness you chose?

LESSON REVIEW

Review Concepts

1. **Explain** the health benefits of moderate- and vigorous-intensity activities.
2. **Describe** how aerobic and anaerobic exercises affect your body.
3. **Explain** how to work out safely and effectively.
4. **List** factors you should consider when you set health goals for fitness.

Critical Thinking

5. **Analyze** Why is frequency important when choosing exercise to build your strength and endurance?
6. LIFE SKILLS **Set Health Goals** Choose one fitness skill from Lesson 3. Explain how improving that skill would affect your health.

LESSON 5

Play It Safe

You can reduce the risk of injury by following safety rules. You also can learn what to do if someone is injured.

You will learn . . .

- what safety equipment and safety rules can help protect you from injury.
- how to prevent and treat injuries before, during, and after a workout.
- how to be a good sport.

Vocabulary

- **safety equipment standards**, *C31*
- **muscle strain**, *C33*
- **sprain**, *C33*
- **PRICE treatment**, *C33*

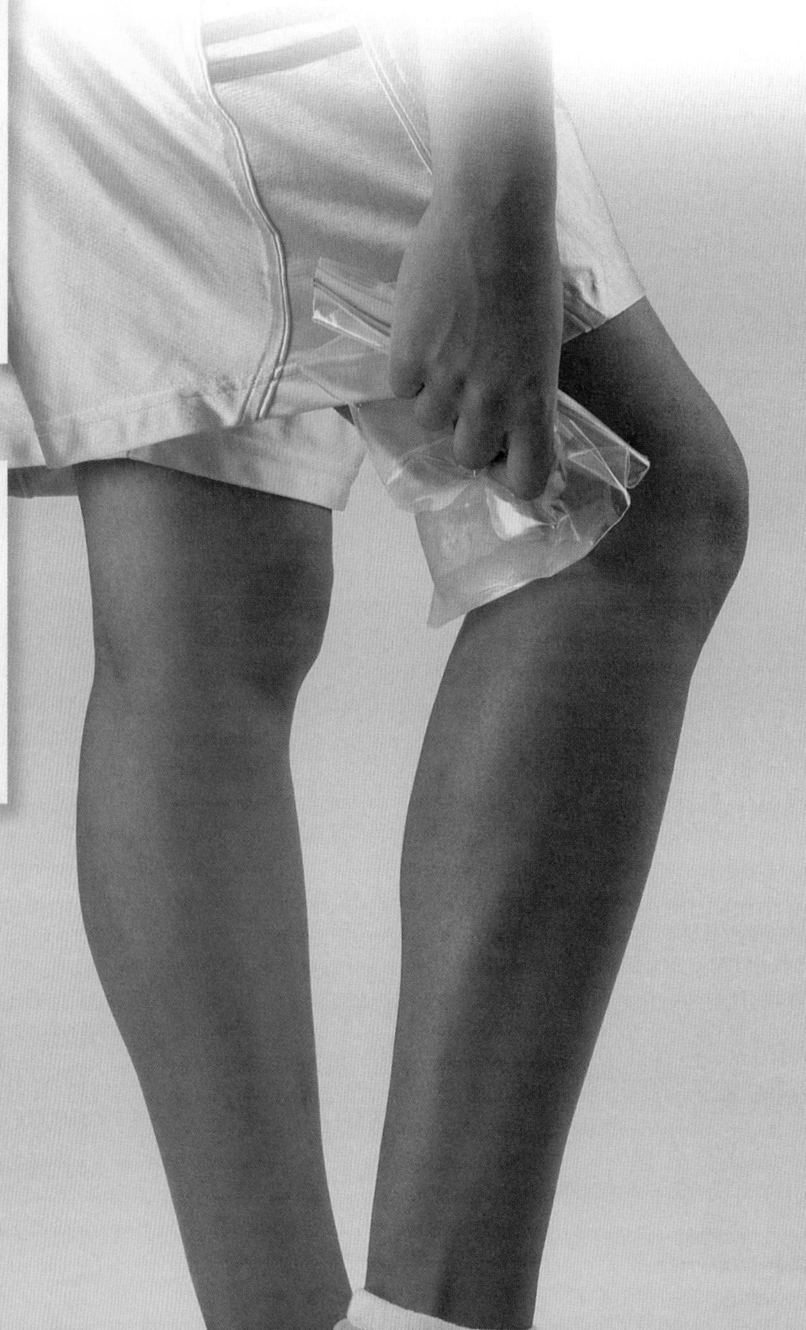

Digital Light Source - Richard Hutchings/McGraw-Hill Education

Safety in Sports and Games

Safety equipment is designed to protect you when you play sports and games. Some activities need more safety equipment than others. For example, if you play football, you must wear a helmet with a face mask; pads on your shoulders, chest, arms, and legs; a mouth guard; and shoes with cleats. When you play basketball, you may only wear safety glasses. Check with your parent, guardian, or sports coach to find out what equipment you need.

Safety equipment should meet basic safety standards. **Safety equipment standards** are rules that say what equipment must be made of and how it must work. Most safety equipment will have a label saying what standards it meets.

Safety equipment should also fit you. If your safety equipment is too small or too large, it may not protect you well. Make sure that you wear your safety equipment the way it's designed to be worn.

Sports and games have safety rules. It is important to follow these rules. They make the games fun and fair for everyone. They also help protect you from injury. No matter what game or sport you play, follow the safety rules for the game.

Health Online

Gear Up for Safety

Choose a sport. Every sport or game requires its own safety equipment. Research the equipment that the sport requires. Report what you learned using the e-Journal writing tool.

Why should you check the fit of safety equipment?

▲ When you play a sport like basketball, follow the rules.

Make a Workout Safety Poster

Make a poster to show the steps for a safe workout. Include drawings or pictures of safety equipment and of simple stretches students your age can do to warm up and cool down. Put your poster in your classroom or school gym to remind other students how to work out safely. With your teacher's permission, demonstrate the stretches.

A Safe Workout

There are several things you can do before, during, and after physical activity to reduce your risk of injury.

Before Physical Activity

- **Choose which activity** you will be doing.
- **Choose the proper clothing** for the activity. Dress so you will be comfortable. Wear the right kind of shoes.
- **Choose the proper safety equipment** for the activity. For example, wear a helmet while bicycling and knee pads when skateboarding.
- **Make sure all your equipment** is in good working order. For example, check the brakes on your bicycle before you ride. Doing this helps you stay safe.
- **Warm up** to prevent muscle injuries.
- **Drink water** to prepare for the water your body will lose when you perspire.

During Physical Activity

- **Drink more water** to replace the water your body loses as you sweat.
- **Keep track** of your heart rate. Is it in your target zone?
- **Rest** if you need to.
- **Stop exercising immediately** if you think you have an injury.

After Physical Activity

- **Cool down** to keep your muscles loose.
- **Check your heart rate** to make sure it returns to normal.
- **Continue to drink water** to replace the water lost through perspiration.

Strains and Sprains

Injuries can occur even when you are careful. Two common injuries are muscle strain and sprains. A **muscle strain** is an overstretch of a muscle. A strained muscle can hurt a lot. You can strain a muscle if you don't warm up. People who don't have good flexibility are more likely to strain muscles.

A **sprain** is an injury to the tissue that connects bones to a joint. You can prevent sprains and muscle strains by learning how to do your activities properly. Practice helps, too. The boxes below show how to treat muscle strains and sprains if they do happen.

What can you do to prevent sprains and muscle strains?

Muscle Strain

- Apply ice for 15 minutes every three to four hours for the first day.
- Keep the strained muscle elevated.
- Rest the muscle for at least a day.
- Avoid using the strained muscle while it is still painful.
- When the muscle starts to feel better, begin using it again. Use it gently at first.

Sprain

The **PRICE treatment** for sprains and other injuries to muscles and bones includes five steps.

Protect Use an elastic bandage, splint, or sling to protect the injured area.

Rest Stop using the injured arm or leg right away.

Ice Put ice on the area.

Compression Wrap the injured area in a soft bandage.

Elevation Raise the injured area to a level above the heart.

CAREERS

Personal Trainer

A personal trainer helps people train to improve or maintain health fitness and fitness skills. The personal trainer may help a person plan a workout and keep following it. Personal trainers often work with professional athletes. They may work with people who want to begin a fitness program. Personal trainers can show people how to do exercises correctly and safely. A personal trainer might help someone rehabilitate a serious injury.

Be a Good Sport

Perform a Skit

Respect How can you show respect for everyone involved in a game? With a partner, perform a skit showing how to show respect when your team wins and when your team loses.

Do you play team sports or take part in other athletic competitions? Do you enjoy watching other people compete? When everyone tries to be a good sport during competitive events, the game is more fun! A *good sport* plays fair, respects others, and follows the rules during sports and games.

- **Compete with respect.** When you compete, you want to win. That's normal. Winning can make you feel great. When you are a good sport, you try to win, but you respect your teammates and members of the opposing team at the same time. You do not tease, shove, hit, kick, or fight with other players. You follow the rules of the game and show respect to everyone involved.
- **Lose with grace.** If you lose a competition, keep a positive attitude. Pat yourself on the back for the hard work and effort you put into the game. Your coach can help you and your teammates work together to improve your skills. He or she can also help you practice good sportsmanship and keep a positive attitude.
- **Be a good fan.** Fans can also be good sports. Fans are like an audience for sports and competitions. Your parents or guardian, other family members, and friends may come to watch you play. Good sports cheer for their team whether the team wins or loses. Their support can help you stay positive during a tough game.

Cheering for your favorite team is fun. It helps the players do well, too. How can you support your team and still be a good sport?

Sometimes playing in front of people may make you feel nervous. This is a good time to give yourself a pep talk. Picture yourself playing well. Don't put yourself down if you make a mistake.

It is important that fans and observers be good sports as well. If you are watching your friend play basketball, cheer her on. Do not tease or insult other players, even if they make mistakes.

Participating in sports and games should be enjoyable for the coach, the players, and the fans. You should not feel pressured to play perfectly all the time. If you feel unhappy or stressed about a sport you play, talk to your parents or guardian or another responsible adult.

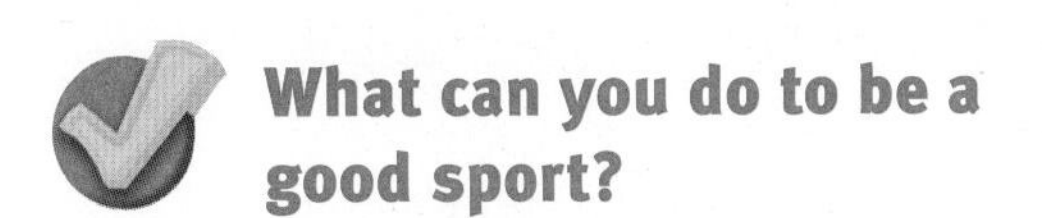

What can you do to be a good sport?

ACTIVITY

LIFE SKILLS

CRITICAL THINKING

Use Communication Skills

With a group, role-play healthful ways to communicate your feelings when you win or lose a game. Divide into two "teams." Decide who won and who lost.

1. **Choose the best way to communicate.** Think about ways you have seen people communicate at a sports event or on TV. Have the team who lost choose a healthful way to communicate their feelings.
2. **Send a clear message. Be polite.** The team who lost should send a clear message to the team that won.
3. **Listen to each other.** The team that won will listen. Then that team will respond as the team that lost listens.
4. **Make sure you understand each other.** In your role-playing, make sure people on each team understand what the other team said.

LESSON REVIEW

Review Concepts

1. **List** three things you should check when you use safety equipment.
2. **Identify** three things you can do before, during, and after a work out to prevent injuries.
3. **Discuss** ways to be a good sport.

Critical Thinking

4. **Compare** What should you do to treat both a sprain and a muscle strain?
5. LIFE SKILLS **Use Communication Skills** Suppose that a teammate makes a very good play. How can you celebrate and still be respectful to the other team?
6. LIFE SKILLS **Access Health Products** Why is it important to know about safety equipment standards when you and your parents or guardian choose safety equipment?

Learning LIFE SKILLS

CRITICAL THINKING

Access Health Facts, Products, and Services

Problem You need to buy a bicycle helmet. There are many different helmets in the store. How can you and your parent or guardian decide which one to choose?

Solution You can collect information to help you and your parent or guardian decide. You can find out which helmet meets safety standards and fits you best.

Learn This Life Skill

Follow these steps to access health information. The Foldables™ can help you.

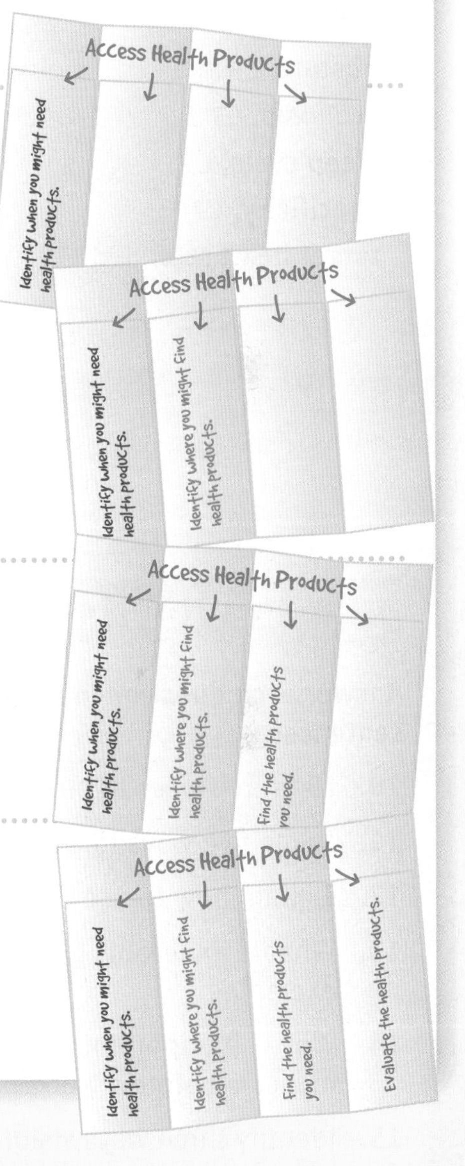

1 Identify when you might need health products.

When do you need to wear a bicycle helmet?

2 Identify where you might find health products.

Where can you and your parent or guardian buy a bicycle helmet? Where can you find information about safety standards for helmets? Can you ask a salesperson? Can you do research at the library or on the Internet?

3 Find the health products you need.

Go to the sources you chose. Write down the information you learn. Then go to a place that has the health product you need. You could go to a bicycle store.

4 Evaluate the health products.

Based on the information you found, which helmet would you and your parent or guardian buy? Why? Where would you buy it? Explain.

Practice This Life Skill

Identify the safety equipment you need to play another sport. List the information you need to know to choose the best product. With your parents or guardian, go to a store or look at a catalog. Then choose the best equipment for you and explain why you chose it.

CHAPTER 5 REVIEW

Use Vocabulary

cardiac output, *C18*
decibel, *C15*
overload, *C26*
periodontal disease, *C11*
posture, *C5*
scoliosis, *C5*
sleep cycle, *C9*
specificity, *C26*
target heart rate, *C26*

Choose the correct term to complete each sentence.

1. A curving of the spine to one side of the body is called __?__.

2. The amount of blood pumped by your heart is your __?__.

3. A(n) __?__ is a unit used to measure the loudness of sound.

4. Working out to improve particular areas of fitness is __?__.

5. When you work your body harder than normal, you __?__.

6. The stages your body goes through as you sleep make up the __?__.

7. A disease of the gums and the bone that support the teeth is __?__.

8. A fast and safe heart rate for workouts is your __?__.

9. The way you hold your body as you sit, stand, and move is your __?__.

Review Concepts

Answer each question in complete sentences.

10. List three ways to protect your ears.

11. Name five ways to care for your eyes.

12. Explain how to use the PRICE treatment.

13. What is anaerobic exercise? Give three examples.

14. Why is it important to warm up your muscles before a workout?

15. Identify three ways vision problems can be corrected.

16. Explain what happens during a medical checkup.

Reading Comprehension

Answer each question in complete sentences.

If you lose a competition, keep a positive attitude. Pat yourself on the back for the hard work and effort you put into the game. Your coach can help you and your teammates work together to improve your skills for the next time. He or she can also help you practice good sportsmanship and keep a positive attitude.

17. What is a good thing to do if you lose a competition?

18. How can your coach help you and your teammates show good sportsmanship?

19. Give an example of how you can be a good sport when you lose a game.

Critical Thinking/Problem Solving

Answer each question in complete sentences.

Analyze Concepts

20. Explain how you can be responsible for your own safety when participating in physical activities.

21. Why is it important to get enough sleep?

22. Why should you not put objects into your ears? Explain what might happen if you did.

23. Explain why getting physical activity is a healthful action.

24. Why should a person participate in both aerobic and anaerobic exercise?

25. By brushing your teeth and flossing daily, how are you protecting your teeth?

Practice Life Skills

26. **Access Health Products** You want to buy a pedometer, an instrument that measures how far you walk or run. List at least four sources you could use to get the information you need to buy a pedometer.

27. **Make Responsible Decisions** You have a piano recital tomorrow. You know you need enough sleep to do well. Your best friend invites you to a sleepover tonight. What should you do? Use the *Guidelines for Making Responsible Decisions*™ to help you decide.

Read Graphics

Use the graph to answer the questions.

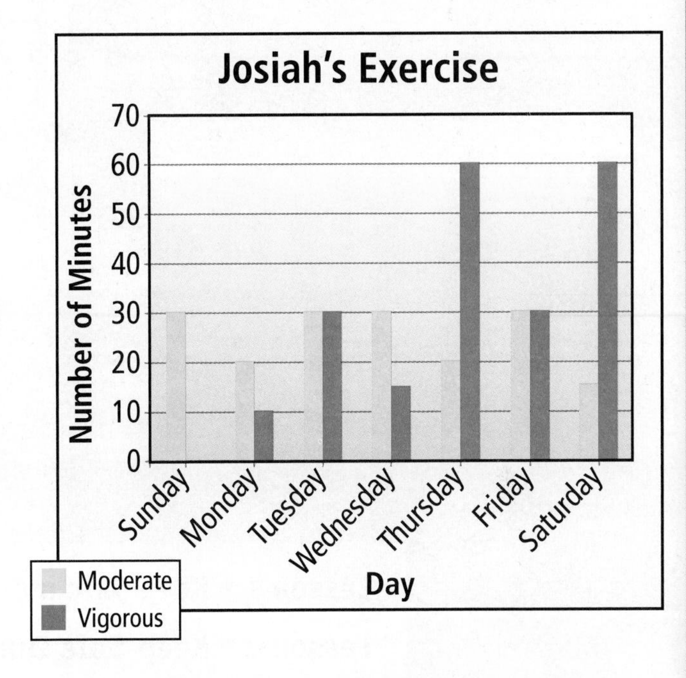

28. On which days did Josiah get at least 30 minutes of moderate-intensity physical activity?

29. On which days did Josiah get at least 30 minutes of moderate-intensity physical activity and 30 minutes of vigorous-intensity physical activity?

30. How could Josiah improve his exercise plan?

CHAPTER 6

Violence and Injury Prevention

Lesson 1 • Keep Safe Indoors *C42*

Lesson 2 • Keep Safe Outdoors *C48*

Lesson 3 • How to Handle Emergencies *C54*

Lesson 4 • Facts on First Aid *C60*

Lesson 5 • Staying Violence Free *C66*

Lesson 6 • Steering Clear of Gangs *C74*

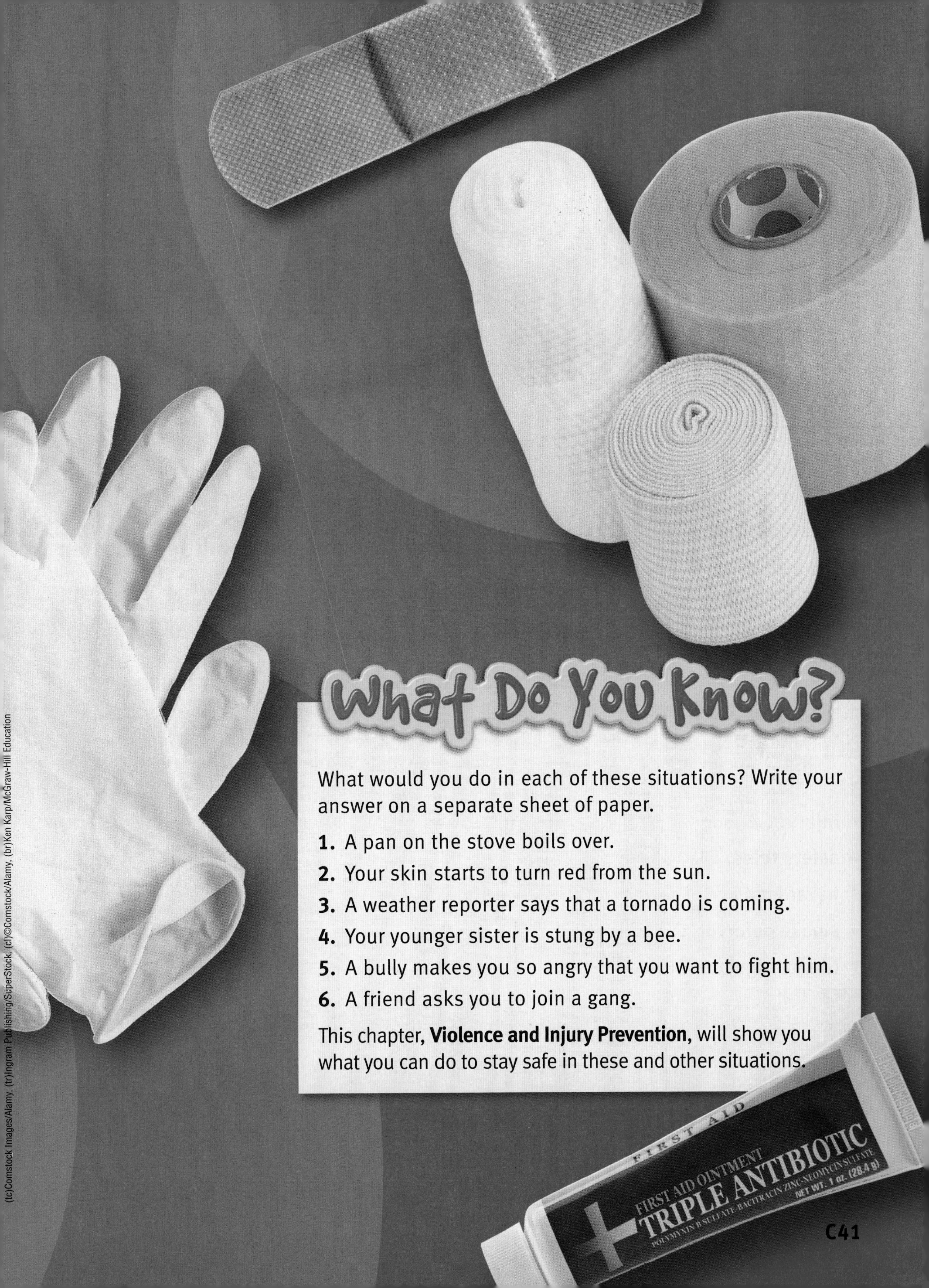

What Do You Know?

What would you do in each of these situations? Write your answer on a separate sheet of paper.

1. A pan on the stove boils over.
2. Your skin starts to turn red from the sun.
3. A weather reporter says that a tornado is coming.
4. Your younger sister is stung by a bee.
5. A bully makes you so angry that you want to fight him.
6. A friend asks you to join a gang.

This chapter, **Violence and Injury Prevention**, will show you what you can do to stay safe in these and other situations.

LESSON 1

Keep Safe Indoors

You will learn . . .

- how to reduce safety hazards at home.
- what safety rules to follow in case of fire.
- how to stay safe at school.

Vocabulary

- **injury**, *C43*
- **safety rules**, *C43*
- **hazard**, *C43*
- **smoke detector**, *C44*
- **fire extinguisher**, *C44*

Every year about two-and-a-half million children visit the hospital due to a fall. Another 200,000 children are injured on playgrounds. You can help reduce the risk of accidents at home and in school.

Safety at Home

Many accidents occur in the home. Accidents can cause injury. An **injury** is harm done to a person. Some accidents cause minor injuries, such as a scraped knee or a small cut. Others can cause more serious injuries. These injuries can take a long time to heal. The person who is injured may need to stay home from school or work. He or she may need extra care.

It's important to be safe at home. You can reduce the risk of injury by following safety rules. **Safety rules** are guidelines to help prevent injury. Many safety rules help you avoid hazards. A **hazard** is something that can cause harm or injury. Here are some ways you and your family can reduce hazards in your home.

- **Keep floors clear of objects** so that people don't trip and fall.
- **Keep cleaners and medicines in separate cabinets** where small children can't reach them. These products can be poisonous.
- **Use electrical items carefully.** Follow the directions. Don't use them when you are wet. Use power tools only with a responsible adult.
- **Store medicines in a locked cabinet.** This will keep them out of the reach of small children.

▲ Put tools away safely.

ACTIVITY

Science LINK

Draw a Diagram

Carbon monoxide is a gas that can make people ill. Any appliance, such as a furnace or stove, that burns gas, oil, coal, or wood can release carbon monoxide if it does not work properly. In most cases, this is not a problem if there is enough fresh air. If there is not enough fresh air, the carbon monoxide can poison people. Many homes now have carbon monoxide detectors. These detectors set off an alarm if the level of carbon monoxide in the air becomes dangerous. Draw a diagram of your home showing where you might put these detectors.

What is the safe way to store cleaners and medicines?

Fire Safety

CAREERS

Firefighter

A firefighter is trained to perform many kinds of rescue missions. He or she fights fires and rescues people who may be trapped in a burning building. Firefighters also respond to emergency medical calls and accidents. In fact, firefighters are often the first people to arrive when there is an auto accident or someone needs an ambulance. Firefighters also help teach people about fire prevention and safety.

Fires and burns injure almost 500,000 people every year. Many fires start by accident. Maybe a pot of food on the stove or an electrical appliance catches fire.

Smoke detectors and fire extinguishers help protect your home. A **smoke detector** is a device that sounds an alarm when smoke is present. A **fire extinguisher** is a device containing water or chemicals to spray on a fire.

Work with your family to plan what to do if there is a fire. Decide where to put smoke detectors and when to test them. Plan how you would get out if there were a fire. Arrange where to meet once you're out of the house.

Fire Prevention Tips

Many fires can be prevented. You can help. Here are some safety rules to remember.

- **Turn pan handles toward the center of the stove.** This way you will be less likely to knock the pan over or touch the hot handle.
- **Use potholders when you move a hot pan.** This protects your hands from burns.
- **Never play with matches.** You could accidentally start a fire.
- **Make sure that electrical cords don't run under rugs.** If the cord is damaged, the electricity can set fire to the rug.
- **Don't plug too many appliances into one outlet.** Doing so can overload an outlet. This can cause a fire.
- **Don't leave appliances running with no one nearby.** Someone should watch the appliance to make sure it does not catch fire.

In Case of Fire

If a fire breaks out, you will need to think and act fast. You can prepare yourself by learning what to do. Ask your family to practice what to do by having fire drills at home. Pay attention when you have fire drills at school, too. Have a plan to get out of your home or school if there is a fire. You should know two ways to get out in case one way is blocked.

Here are steps to follow if there is a fire at home.

- **Yell loudly** to alert others at home. Yell "Fire!" and "Get out!"
- **Don't stop** to take anything. It's more important to protect your safety.
- **Feel whether a door is hot** before you open it. If it feels hot, don't open it. There may be a fire on the other side. If the door is hot, put blankets or clothes along the bottom of the door to keep smoke out. Open a window and yell for help. Wait for firefighters to help you.
- **Be sure to close the door behind you** if you decide it is safe to leave. This will slow the spread of the fire.
- **Crawl on your hands and knees** if there is smoke. Smoke usually rises. Stay below the level of the smoke so that you don't breathe it in.
- **Meet your family outside.** Use the emergency plan you practiced. Call for emergency help after you get out of the building.

If a fire breaks out at school, stay quiet. Pay attention to your teacher's instructions. Use the escape plan that you practiced. Don't run. Go to a meeting place that your teacher and school have decided on. That way people will know that you are safe.

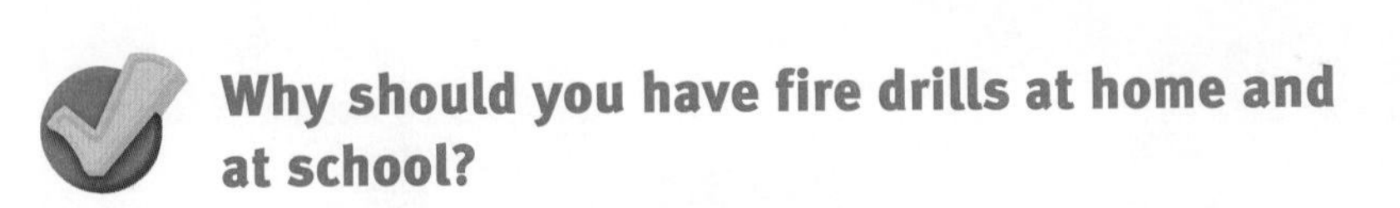

Why should you have fire drills at home and at school?

On a poster board, draw a floor plan of your home. Label all rooms and include a family meeting place outside. Draw arrows showing how you would get out of the home if there was a fire. Show two different ways to get out from your bedroom.

▲ **Put smoke detectors on every level of your home. One should be placed near every bedroom. Test the batteries regularly. Change dead batteries right away.**

Safety at School

Your principals and teachers work to keep your school safe. You can help, too. Here are some ways to stay safe at school.

- **Keep the hallways and aisles in your school clear** so that people don't trip.
- **Don't run** in hallways. It is easy for someone to get knocked over.
- **Don't push anyone else,** especially people who are drinking from water fountains. If you push someone drinking from a fountain, the person could hit his or her head on the fountain.
- **Tell a teacher** if you see a wet floor. Someone could slip and fall.

Use only playground equipment that is safe and not broken. Check the ground under anything you climb on. Make sure that it has a thick layer of soft material, such as sand. Don't play near a construction site or near broken glass.

Bananastock/age Fotostock

▲ **Don't push or shove people in lines.**

Always follow safety rules for sports and games. Don't push or crowd other people. Always play with a friend when you play outside. Tell a parent or guardian where you will be playing. These two tips make sure that someone can help you if there is a problem.

What should you look for when you play on a playground?

ACTIVITY LIFE SKILLS

CRITICAL THINKING

Set Health Goals

Playing safely can help you avoid injury. You can set a goal to follow safety rules.

1. **Write the health goal you want to set:** I will follow safety rules when I play. Write your goal in a Health Behavior Contract.
2. **Explain how your goal might affect your health.** How will playing safely keep you healthy? Write your answer in your contract.
3. **Describe a plan you will follow to reach your goal. Keep track of your progress.** List five safety rules you can follow when you play. Write them into your Health Behavior Contract. Each time you follow a rule, put a check mark next to that rule.
4. **Evaluate how your plan worked.** At the end of the week, look at your chart. Which rules did you follow regularly? Which rules do you need to practice more? Do you need to make a new plan?

LESSON REVIEW

Review Concepts

1. **List** three ways to prevent injuries at home.
2. **Explain** how to reduce the risk of fire in your home and what to do if a fire happens.
3. **Identify** three ways to prevent hazards at school.

Critical Thinking

4. **Analyze** Explain why you should follow safety rules at school.
5. LIFE SKILLS **Set Health Goals** Set a goal to do three things to remove hazards in your home. Write a Health Behavior Contract stating your plan.

LESSON 2

Keep Safe Outdoors

You will learn . . .

- what safety precautions pedestrians should follow.
- what safety rules to follow when you ride in a car or bus or ride bicycles, scooters, or skateboards.
- what safety rules to follow in severe weather conditions and in the water.

Vocabulary

- **pedestrian**, *C49*
- **seat belt**, *C50*
- **frostbite**, *C52*
- **heatstroke**, *C52*

Do you like to spend time outdoors? Do you like to walk in your city or town? There are many fun things to do outside. There can also be dangers. Following safety rules, such as waiting at crosswalks, can help keep you safe as you enjoy these activities.

Walk Safely

Have you ever walked to school or to a friend's house? A **pedestrian** (puh•DES•tree•uhn) is a person who walks on the sidewalk or in the street. Here are some safety rules for pedestrians.

- **Use sidewalks and crosswalks.** Don't enter the street between parked cars.
- **Obey traffic signals.** Wait for the "Walk" sign. Then look left, right, and left again before you cross.
- **Walk facing traffic if there are no sidewalks.** That way you can see cars that are coming toward you.
- **Wear light-colored or reflective clothes if you are out at dusk or at night.** This makes it easier for drivers to see you.
- **Walk with a friend or responsible adult.** Don't hitchhike.

▲ **How are these people being safe pedestrians?**

Stranger Danger

One important safety rule for pedestrians is to be careful around people you don't know. A stranger might ask you for directions or to help him or her find a lost pet. A responsible adult would ask another adult for help, not someone your age. If someone you don't know asks you for help or invites you into a car, say "no" in a loud, firm voice. Then get away as fast as you can. Run in the direction opposite from the one the car is traveling in. Tell your parent, guardian, or a responsible adult what happened.

Explain why you need to pay attention when you are a pedestrian.

MAKE a Difference

Students at Skinner Middle School in Colorado decided to do something about pedestrian safety near their school. They worked with the police, other schools, and community leaders. They took pictures of traffic. They drew maps showing traffic patterns. They pressured their community to put a stop sign at a dangerous intersection. Can you think of any ways to help make a community safer?

Safety on the Road

You may ride in cars and buses often. You may also skate or ride a bicycle, scooter, or skateboard. It's important to follow safety rules when you are on the road.

Write About It!

Write a Persuasive Essay
Suppose a friend doesn't wear a seat belt. He says it is uncomfortable. Write an essay to persuade him to wear a seat belt. Explain how the belt can keep him safe.

In a Car

Always wear a seat belt. A **seat belt** is the lap belt and shoulder belt worn in a car. It keeps you from being thrown out of the car in a crash. People age 12 or younger should ride in the back seat. This way they won't be injured by the air bags. Air bags inflate when a car hits another object. They prevent a person from hitting the dashboard of the car. But air bags are designed to protect adults. The force of the air bags can harm smaller people.

▲ **Wear a seat belt every time you are in the car.**

Keep doors locked while the car is moving so that they don't open accidentally. When the car has stopped, get out on the side of the car next to the curb. This way you won't step out into traffic.

On the Bus

Whenever you ride the bus, sit still and don't yell or fight. Be ready to get off when the bus is at your stop.

When you get off the bus, hold the handrail so that you don't trip on the stairs. Cross the street in front of the bus, not behind it. Check to make sure the driver sees you before you cross.

Bikes, Skates, and Scooters

Bicycles, scooters, skates, and skateboards are fun to ride and good exercise. Wear a helmet for all these activities. When you skate or use a skateboard, wear elbow and knee pads and wrist guards along with your helmet. When you ride a scooter, wear elbow and knee pads with the helmet.

Always ride on a path and avoid holes and bumps. Ride during the day. Wear bright clothing so drivers can see you. When you ride a bicycle, use hand signals before you turn or stop.

List two safety rules each for riding in a car and on a bus.

CRITICAL THINKING

Be a Health Advocate

1. **Choose a healthful action to communicate.** It is important to wear safety equipment when you ride a bicycle, scooter, or skateboard and when you skate.
2. **Collect information about the action.** How can safety equipment protect you from injury? What injuries can the equipment help you avoid? Test your ideas out on a friend.
3. **Decide how to communicate this information.** Plan a cartoon to help convince your friends to wear safety equipment.
4. **Communicate your message to others.** Draw your cartoon. Post it in your classroom for your classmates to see.

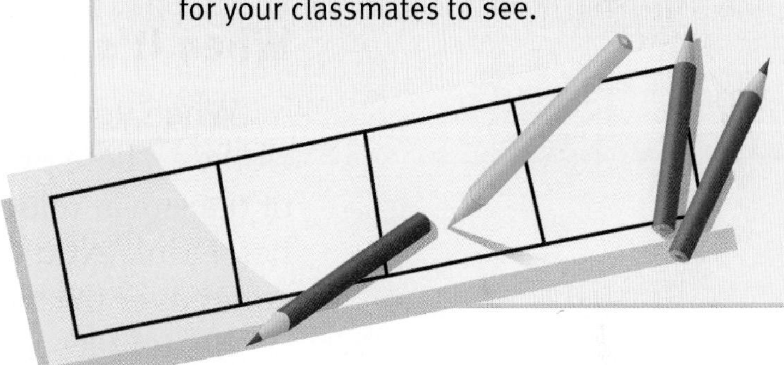

▼ **Signal with your left hand when you ride.**

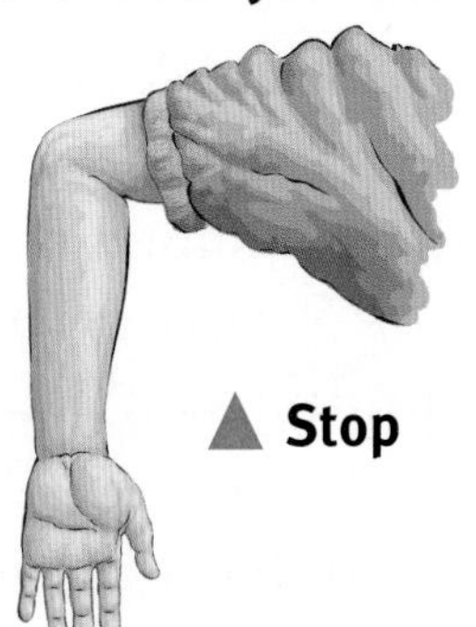

▲ **Stop**

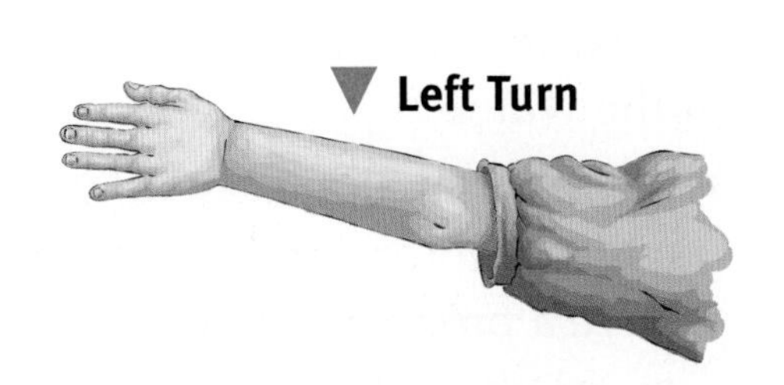

▼ **Left Turn**

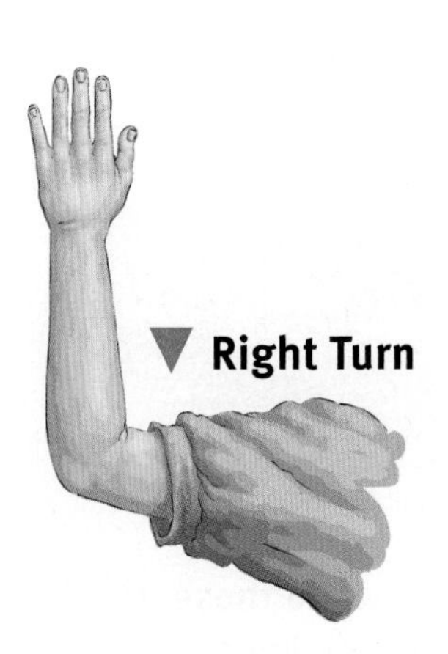

▼ **Right Turn**

Weather and Water

Up to two-fifths of your body's heat is lost through your head. A hat can help keep that heat in.

When It's Cold

The right clothes can help protect you from the cold. A good strategy is to wear several layers of clothes. Wear a hat, a scarf, mittens or gloves, a coat, and shoes or boots that keep out water. Keep moving. Moving your body produces heat.

Peope who don't take precautions in extremely cold weather can be injured. **Frostbite** is an injury caused by exposure to extreme cold. Symptoms include numbness and white, gray, or yellow skin. If you see signs of frostbite, get the person into a warm room. Get medical help right away.

When It's Hot

When you go out in hot weather, wear loose, protective clothes. These protect you from the sun's rays. Stay out of the sun at midday. Wear sunscreen, sunglasses, and a hat. Drink extra water. If you do get sunburned, run cool water over the burn. Don't break blisters.

Heatstroke is harm to the body due to being exposed to high temperatures. Symptoms include hot, dry skin and a very high temperature. If you have these symptoms, get into the shade right away. Sprinkle water on your body to cool it.

▲ **How are these people protecting themselves from the heat?**

In the Water

Learning to swim is the best water safety precaution you can take. That way you will feel more comfortable in the water. Here are some additional tips for staying safe in the water.

- **Never swim alone.** Swim with a buddy. Be sure there is a lifeguard or another responsible adult watching. If you have a problem while swimming, the adult can get help.
- **Don't run or push other people.** You could knock a person who can't swim into the water.
- **Never dive into any lake or river.** The water may be shallow or have rocks hidden beneath the surface.

If you see someone in trouble in the water, call for help. Throw an object that floats to the person, or stick out a rope, a pole, or a branch from a tree. Don't go into the water or try to rescue the person by yourself. Instead, get a lifeguard or have someone call 9-1-1.

What should you do to stay safe in hot weather?

▼ **If you fall in the water, these two methods can keep you afloat until help arrives.**

Back Floating

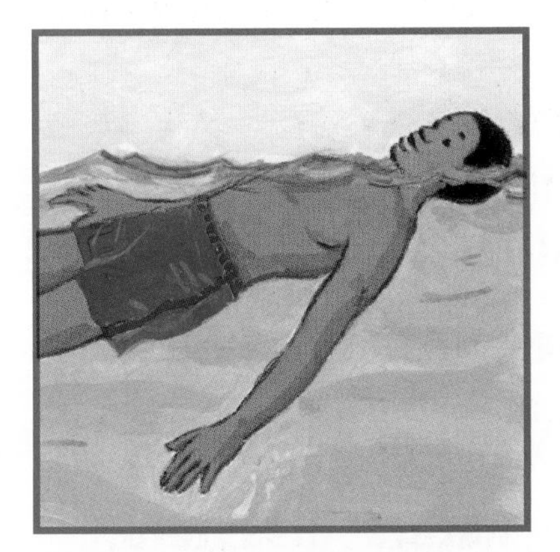

Lie flat on your back. Keep your head above water. Move your arms slowly from side to side. Relax your legs. Kick slightly.

Survival Floating

Inhale. Curl up by tucking your knees into your chest.

Bring your head above water to breathe. Stay above water and breathe for a few minutes. Curl up again to rest.

LESSON REVIEW

Review Concepts

1. **List** three safety rules for walking.
2. **Explain** how to stay safe when you ride in a car or bus or use skates, a skateboard, a bicycle, or a scooter.
3. **Describe** how to stay safe in cold weather, in hot weather, and in the water.

Critical Thinking

4. **Evaluate** You can choose between two places to swim. One is shallow, but there are no lifeguards. The other is deep, but there are lifeguards. Which would you choose? Why?
5. LIFE SKILLS **Be a Health Advocate** Your brother doesn't like to wear a helmet when he rides his bicycle. How could you be a health advocate for your brother?

LESSON 3

How to Handle Emergencies

You will learn . . .

- whom to call in case of emergency.
- how to prepare in case of an emergency.
- what to do to help stay safe in a natural disaster.

Vocabulary

- **emergency**, *C55*
- **hurricane**, *C58*
- **flood**, *C59*
- **earthquake**, *C59*

Suppose that severe weather is coming. Are you and your family prepared? In this lesson you will read about what you can do to keep safe in bad weather and during natural disasters.

What Is an Emergency?

Suppose you get a small cut. You can probably take care of it. But suppose you get a deep cut. It won't stop bleeding. You need emergency medical help. An **emergency** is a situation in which help is needed quickly.

There are many kinds of emergencies. There might be a car accident or a fire. Someone might fall. A person might have a heart attack. A bad storm might blow down power lines. Homeland security might be threatened by a terrorist attack. A *terrorist* is a person who uses violence to try to make another person do what he or she wants.

All these emergencies can be scary. Being prepared for an emergency can help. When there is an emergency, it's important to know whom to call and what to say. The chart below lists some examples.

Role-Play

With your parent or guardian, role-play what to say on the telephone when there is an emergency. Practice each type of emergency. Take turns being the person making the call and the person answering the call.

Whom to Call in Case of an Emergency

Type of Emergency	Who to Call	What to Say
Fire	9-1-1 or your local emergency number for fire	Tell where the fire is and whether anyone is trapped inside.
Serious Injury or Medical Problem	9-1-1 or your local emergency number for medical emergencies	Tell where you are. Describe as much about the problem as you can.
Criminal Acts, Violence	Police or 9-1-1 if the crime is taking place now	Tell where you are and what happened.
Poisoning	Your local poison control number. You can find this number in a phone book.	Tell where you are and what happened. If you have the container that held the poison, tell what the product is.

What should you say when you call about a serious injury?

Planning Ahead

What weather emergencies might happen where you live? Choose one or two. Research how your community prepares for these emergencies. Write a report on your community's emergency plans using the e-Journal writing tool.

You can't tell when an emergency will happen. You can be prepared, though. Knowing who to call is one way to be prepared. You can also prepare in other ways.

Make a Plan

Work with your family to plan what to do if an emergency occurs. Find out what weather emergencies are likely in your community. The Red Cross and other groups can help you find this information.

Many communities have sirens or alarms that warn residents of weather emergencies. If your community has these signals, learn what they sound like. Television and radio stations broadcast emergency signals and information as well. The stations announce the kind of emergency and instructions on what to do.

For any kind of emergency, your family plan should include the following.

Phone Numbers Make a list of important phone numbers that you might need in an emergency. Include emergency phone numbers, your parents' or guardian's work phone numbers, and the phone numbers of other family members. This way you will be able to reach each other if an emergency occurs.

Disaster Kit On the next page, you can read more about what to put in a disaster kit.

Escape Plan Know how to get out of your house in an emergency.

Meeting Place Choose a place to meet near your house. Also have a place to meet that is farther away, in case you can't get back home. Everyone should know where these places are.

Practice your plan with your family so that you will be ready if an emergency occurs.

Make a Disaster Kit

In some emergencies, such as storms, there may be a loss of electricity. You may not be able to get to a store. To help prepare for these emergencies, you can make a disaster kit. Gather the supplies you will need. Keep them in a safe place. Check your supplies every six months.

The supplies you need depend on the kinds of weather emergencies that occur in your area. The checklist below shows some items you might put in your kit.

List four things that should be part of a family emergency plan.

Disaster Kit

- First aid kit: Most first aid kits have several kinds of bandages, pain medicines, and medicines for cuts and other minor injuries.
- Flashlight
- Matches in a waterproof container
- Battery-operated radio
- Extra batteries
- Canned or packaged food and a can opener
- Bottled water
- Maps of your area
- Extra clothes
- Extra blankets and sheets
- Fire extinguisher
- Duct tape and other supplies to repair things that break
- Any medicines your family needs

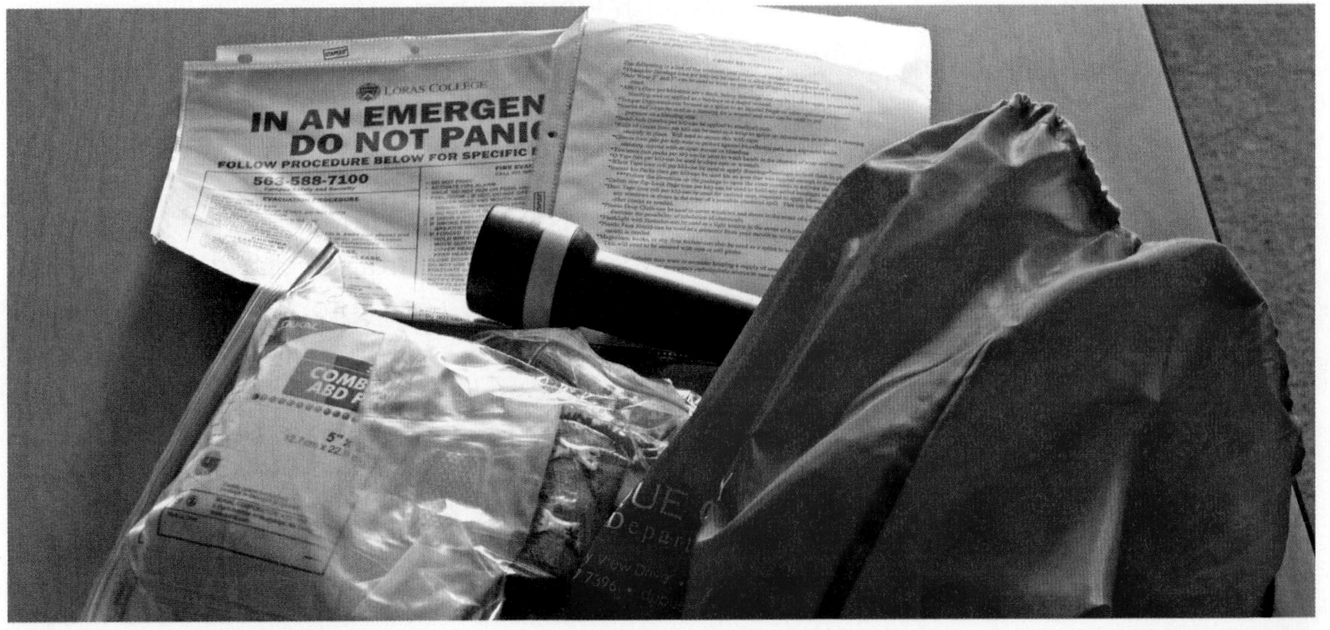

Natural Disasters

▲ Hurricane

▲ Flood

▲ Earthquake

A *natural disaster* is an emergency caused by nature. Storms, floods, and earthquakes are all natural disasters. Learn how to be safe during the kinds of disasters that might occur where you live.

Storms

A *thunderstorm*, or electrical storm, brings heavy rain, thunder, and lightning. If you are outdoors when a thunderstorm strikes, get inside right away. If you can't get inside, squat down and make yourself as small a target as possible. Move away from tall trees. Once you are indoors, stay away from doors and windows. Don't take a bath or shower, talk on the phone, or use electrical objects until the storm passes. Water and electrical wires can carry electricity to you, causing an electric shock.

A **hurricane** is a tropical storm with heavy rain and high winds. A *storm surge* occurs when winds from a storm push water towards the shore. This surge combines with the tide and increases the water level which can cause flooding in coastal areas. If you live in an area where hurricanes occur, have a disaster kit ready. When a hurricane is near, board up windows and doors with wood. When a hurricane hits, stay indoors and away from windows. You may have to evacuate, or leave the area. Sometimes people choose to leave. At other times government officials tell people that they must leave to be safe.

A *tornado* is a fast-moving, funnel-shaped cloud of wind. Tornados can develop during thunderstorms. If you are indoors during a tornado, take shelter in a basement or in a hallway away from windows. Watch out for objects that may be picked up and blown around by the wind. If you are in a car, get out. Find shelter in a building if you can. If you can't, lie flat in a ditch or low-lying area. Cover your head to protect it from flying objects.

Floods and Earthquakes

A **flood** is overflow of water onto normally dry land. Never swim or ride in a car through floodwaters. Leave your home if you are told to do so. A *flash flood* is a sudden flood usually caused by heavy rain. If you are outdoors when a flash flood strikes, climb to higher ground. Don't try to outrun the flood.

An **earthquake** is a shaking or trembling of the ground caused by sudden shifting of Earth's crust. If you are inside during an earthquake, drop to the floor. Get under a sturdy table or desk and hold on to it. Protect your eyes from flying objects. If you don't have a table or desk nearby, sit on the floor. Be sure you are away from bookcases or other furniture that could fall on you. If you are outdoors, try to move away from buildings and electrical wires. Drop to the ground to protect yourself.

What should you do if you are outside during a thunderstorm?

ACTIVITY — LIFE SKILLS — CRITICAL THINKING

Access Health Facts

Broadcast a report about a weather emergency occuring where you live.

1. **Identify when you might need health facts.** Choose a weather emergency that could occur where you live. Write an emergency weather report for your class. Include instructions on what to do if the emergency happens.
2. **Identify where you might find health facts.** Make a list of sources for your report. You could use your library or talk to a weather forecaster.
3. **Find the health facts you need.** Collect the information you need. Write your report. Give your weather report to your class.
4. **Evaluate the health facts.** Have your classmates ask questions. Can you answer them based on the sources you used?

LESSON REVIEW

Review Concepts

1. **Explain** what an emergency is.
2. **Describe** how to create a family emergency plan.
3. **List** ways to stay safe during thunderstorms, hurricanes, tornados, floods, and earthquakes.

Critical Thinking

4. **Classify** List items you would put in a disaster kit. Identify why each item would be important during an emergency.
5. LIFE SKILLS **Access Health Services** How could you find information about agencies in your community that help during disasters?

LESSON 4

Facts on First Aid

You will learn . . .

- how to give first aid for minor injuries.
- what CPR and rescue breathing are.

Vocabulary

- **first aid**, *C61*
- **tetanus**, *C61*
- **universal precautions**, *C61*
- **fracture**, *C62*
- **CPR**, *C63*

There are times when you need emergency medical help right away. At other times you can handle the situation yourself. There are special ways to treat different kinds of injuries.

What Is First Aid?

Many common injuries can be treated with first aid. **First aid** is the quick and temporary care given to a person who has a sudden illness or injury. The chart on the next page will give you tips on how to perform first aid for many common injuries.

Minor Cuts and Bleeding

When someone gets a minor cut, clean it with soap and water. Put pressure on it with a bandage if it bleeds. Then cover the cut with a clean bandage.

A doctor should check for tetanus if you get a puncture wound. A *puncture wound* is a deep cut, such as the cut you would get from a pin, a nail, or an animal bite. **Tetanus** is an infection caused by poisons made by bacteria that enter a puncture wound. This disease attacks the nervous system. To stay safe you should keep your tetanus shots up to date.

Universal Precautions

Universal precautions are steps taken to avoid having contact with pathogens in body fluid.

- Wear disposable gloves.
- Don't wear the same gloves more than once.
- Wash your hands with soap and water after you take off the gloves.
- Don't eat or drink while giving first aid.
- Don't touch your mouth, eyes, or nose while giving first aid.

First Aid for Poison

Household products such as cleaners, paints, and medicines can be poisonous. If you think someone has swallowed or touched a poison, call 9-1-1 or Poison Control. The people at Poison Control know how to treat poisoning in an emergency. Save the container of the poison so that you know what it is.

Science LINK — ACTIVITY

Bandage Test

Find out which bandages work best. Bandages absorb the blood that comes from a cut. Collect several different kinds or brands of bandages and an eyedropper. Use the eyedropper to drop water on to each bandage. Count how many drops each bandage holds before it stops absorbing water. Which holds the most? Which holds the least? Make a chart of your results.

▲ **People at Poison Control may ask you what the person swallowed or touched, and how long ago it occurred.**

Tim Fuller/McGraw-Hill Education

First Aid for Minor Injuries

Type of Injury	Description and Causes	First Aid Steps
Bee sting	A bee will leave its stinger in the skin. The stinger injects venom, or poison into the skin	• Scrape the stinger out with a card edge or nail file. • Clean with soap and water. • Put ice on the area. • Some people are very allergic to stings. If the person feels dizzy or can't breathe, get emergency medical care.
Spider bite	Some spiders, such as the brown recluse and black widow shown here, can be poisonous.	• Find the spider that bit you if possible. If it is a brown recluse or black widow or you aren't sure what kind of spider it was, call 9-1-1. • Wash the area with soap and water, for other bites. Put ice on the bite to stop any swelling.
Blister	A blister is an area under the skin where fluid collects.	• Clean with soap and water. • Cover with a clean bandage. • Don't break the blister.
Minor burn or sunburn	A minor burn can be caused by heat or chemicals. Sunburn is caused by exposure to the sun's harmful rays.	• Place a cold cloth over the burn, or run cold water on it for 10 minutes. • Cover with a clean bandage.
Fracture	A **fracture** is a break in a bone.	• Don't move the injured body part. Put ice on it. • Call 9-1-1.
Nosebleed	A nosebleed is an injury to blood vessels inside the nose.	• Sit down and lean slightly forward. Pinch your nostrils shut for 10 minutes. • Get medical help if it bleeds for more than 10 minutes.
Rashes from plants	Poison ivy, poison oak, and poison sumac can cause skin rashes.	• Run cold water over the rash. • Use calamine lotion to stop itching.

List the first aid steps for a nosebleed.

Rescue Breathing and CPR

There is help for a person who stops breathing and whose heart stops beating. *Rescue breathing* is a way to help another person get air when he or she can't breathe. CPR stands for cardiopulmonary resuscitation (CAHR•dee•oh•POOL•mu•nayr•ee ri•su•si•TAY•shun). **CPR** is a method of reviving someone who has stopped breathing and who has no heartbeat, by mouth-to-mouth breathing and strong rhythmic pressing on the chest.

Only people who have training in rescue breathing or CPR should perform the techniques. You can read about them so you will know how they work.

Write About It!

Write a Public Service Announcement Explain why people should learn CPR. Come up with a catchy slogan, too. Illustrate your announcement. Post it in your classroom or around your school.

Rescue Breathing

Before someone performs rescue breathing, he or she should make sure it is safe to go near the person who is in trouble. Check for exposed electrical wires or danger from traffic. Once it is safe, these are the steps to use.

1. Check to see whether the person is conscious. Tap him or her and shout, "Are you okay?"
2. Call 9-1-1.
3. Lay the victim on his or her back.
4. Gently tip the person's head back. Open his or her mouth. Pull the chin forward.
6. Listen for breathing sounds. Feel for air that the person may be exhaling. See whether the chest is rising and falling.
7. If there is no breathing, pinch the person's nostrils closed. Slowly blow two full breaths of air into the victim's mouth. Some emergency workers use special plastic shields to protect against the person's saliva.
8. Is the person now breathing, moving, or coughing? If not, he or she needs CPR.

▲ **When you are older, you can take a CPR class to learn how to help others.**

CPR

Giving CPR incorrectly can cause serious injury. It should never be used on a person whose heart is beating on its own. Here is how CPR works.

The steps in the basic cycle of CPR can be remembered as "C-A-B" (Chest compressions, Airway, Breathing). The American Heart Association now recommends this sequence for all age groups except newborns.

First, attempt to determine whether the victim is conscious. If the person does not respond, call 911.

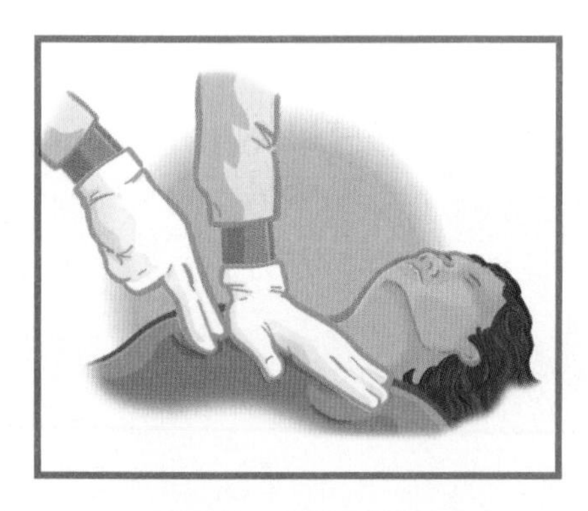

Step 1: If the victim has no pulse, use CPR to restore blood circulation. Kneel beside the victim. Tilt the chin up to open the airway. If the victim is not breathing, do the following: Find the lower part of the breastbone and measure the width of two fingers below that point. Place the heel of your hand on this spot and place your other hand on top.

Step 2: Move so that your shoulders are lined up above your hands. Lock your elbows. Press down, or compress, on the chest 30 times.

What is the purpose of CPR?

ACTIVITY

LIFE SKILLS

CRITICAL THINKING

Make Responsible Decisions

You are playing soccer with your friend when she steps on a nail. She gets a puncture wound. She says that it's nothing, but you are not so sure. Pair up with a partner. Explain to your partner what you would do and why. Then have your partner explain to you what he or she would do.

1. **Identify your choices.** Should you do nothing? Treat the injury yourself? Call 9-1-1? What else could you do? Tell your partner what your choices are.
2. **Evaluate each choice. Use the *Guidelines for Making Responsible Decisions™*.** Tell your partner whether each choice follows the guidelines.
3. **Identify the responsible decision.** Discuss with your partner which choice you would make. Explain why it is a responsible choice.
4. **Evaluate your decision.** Explain to your partner how your decision would affect your friend's health.

Guidelines for Making Responsible Decisions™

- **Is it healthful?**
- **Is it safe?**
- **Does it follow rules and laws?**
- **Does it show respect for myself and others?**
- **Does it follow family guidelines?**
- **Does it show good character?**

LESSON REVIEW

Review Concepts

1. **Define** first aid.
2. **List** the steps of both rescue breathing and CPR.

Critical Thinking

3. **Assess** A friend has blisters on his hands. He doesn't know how he got them. What questions could you ask to help him identify what kind of minor injury it is?
4. LIFE SKILLS **Make Responsible Decisions** You are walking with a friend when she falls and injures her leg. She is afraid that it may be fractured. What should you do? Why is your choice a responsible decision?
5. LIFE SKILLS **Practice Healthful Behaviors** Identify the five universal precautions. Explain how each precaution helps keep you safe when touching someone else's body fluids.

LESSON 5

Staying Violence Free

You will learn . . .

- to identify some signs of violence.
- ways to express anger and resolve conflict without violence.
- where victims of violence and abuse can get help.

Vocabulary

- **weapon**, *C67*
- **discrimination**, *C67*
- **law**, *C70*
- **justice**, *C70*

You may have seen fighting in your community. You may have seen images of war on TV. Unfortunately violence occurs in the world. You can learn to avoid violence. You can get help if violence happens.

The Many Faces of Violence

There are many types of violence. Suppose someone at school threatens to beat you up. This is a kind of violence called bullying. A person might use a gun to threaten another person. A gun is an example of a weapon. A **weapon** is a device used for violence.

Not all violence is physical violence. Calling someone names is a kind of violence. Discrimination can also be a kind of violence. **Discrimination** is treating some people in a way different from how you treat others. Putting down people who have a different race or religion is a kind of discrimination.

If someone touches your private body parts, that is abuse. *Abuse* is another kind of violence. If someone you trust hits you, says mean things, or doesn't take care of you, that is abuse, too.

Violence on TV

List your favorite TV programs. Find out what rating each one has. TV ratings tell whether the program is appropriate for children, teens, adults, or everyone. They are included in the TV listings. The ratings are also shown at the beginning of the TV program and are based on the content of the program. Make a chart showing the ratings of your favorite programs.

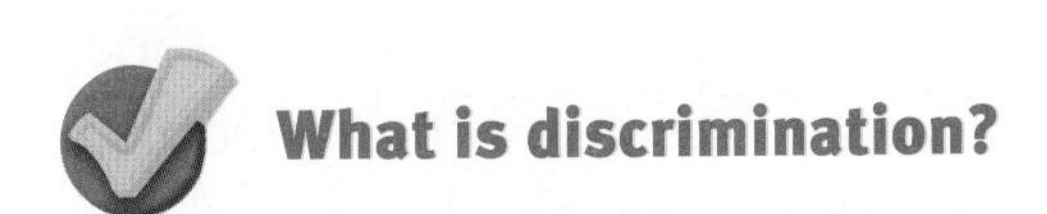

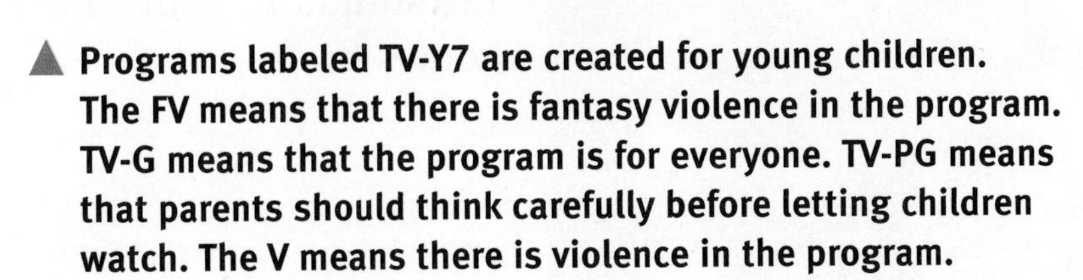

▲ **Programs labeled TV-Y7 are created for young children. The FV means that there is fantasy violence in the program. TV-G means that the program is for everyone. TV-PG means that parents should think carefully before letting children watch. The V means there is violence in the program.**

Dealing with Violence

What if someone wrongly accuses your friend of stealing something? Your friend might become so angry that he or she wants to fight. Your friend's anger is normal. It's okay to be angry. Everyone feels angry sometimes. But anger can lead to violence. What behaviors could help your friend avoid violence?

Manage Anger

Anger management can help you avoid violence. Take time to cool down before you do anything. Think about what you can do to fix the situation that caused anger. What choices do you have? Talk to a parent or guardian to help you decide how to handle your anger. Don't choose violence.

▼ **How can you use communication skills to help avoid fighting?**

Make Wise Choices

You can reduce your risk of violence in other ways, too. Show respect to others. Don't put other people down. This will help you avoid conflicts. If there is a conflict, talk about it instead of fighting. The tips on the next page can help you resolve conflicts.

Choose friends who don't use violence. If your friends choose violence, you may be more likely to be caught in the middle. Don't hang around when people are violent. Get away from the situation. Your safety comes first.

ACTIVITY

LIFE SKILLS

CRITICAL THINKING

Use Communication Skills

With a partner, role-play how to handle a bully. One of you should play the person being bullied. The other should play a teacher or other responsible adult.

1. **Choose the best way to communicate.** Sometimes it is not safe to talk to a bully. In these cases, it is better to talk to an adult who can help.
2. **Send a clear message. Be polite.** The person being bullied should use I-messages to describe how the bullying makes him or her feel.
3. **Listen to each other.** The teacher should use active listening skills. The person being bullied should listen if the teacher responds.
4. **Make sure you understand each other.** Did the teacher "get the message"? See if you can work together to find a solution. Then switch roles and try again.

Someone who carries a weapon increases his or her risk for violence. Even if he or she doesn't use it, someone else might take it and use it. It's safer not to carry a weapon at all.

Be careful around strangers, too. Don't get in a car with someone you don't know. If you use the Internet, don't give anyone your name, address, telephone number, age, or photo. Tell a responsible adult if someone tries to find out your age or where you live.

Use Conflict Resolution Skills

When there is a conflict, use these tips to help resolve it.

- **Calmly discuss** what happened.
- **Be honest** about what you have said or done.
- **Use I-messages** to express your feelings. Say "I'm angry" instead of "You're wrong."
- **Listen to the feelings** of the other person.
- **Discuss** possible solutions. Agree on a solution.
- **Keep your word** and follow the solution.
- **Ask a responsible adult** for help if you cannot agree.

How can you express anger to another person without violence?

Help for Victims of Violence

ACTIVITY

Social Studies LINK

Advocate for Victims

Find out who in your community provides services for victims of violence. With a partner, interview a counselor who works with these people. Record the interview and present it to your class.

A *victim* of violence is a person who has been harmed. It can be painful and scary to be a victim. Victims often have both short-term and long-term effects from the violence. What can they do to get help?

The victim can call for emergency help right away. Police and medical workers will respond. The police will ask the victim what happened and who caused the violence. Emergency medical workers will treat any injuries.

Many violent acts are against the law. A **law** is a rule that people in a community, state, or nation are required to follow. If someone breaks a law, his or her victim can get justice. **Justice** is fairness for all people. Police, detectives, and judges work to help victims get justice.

▲ **Counselors can help victims of violence talk about their feelings.**

Recovering from Violence

Some effects of violence last a long time. The victim may feel afraid, depressed, or ashamed. This is normal, but a victim doesn't have to blame himself or herself. He or she can get help.

One important step toward recovery is for the victim to talk about his or her feelings. Many victims feel angry, ashamed, or sad. They can talk to a parent or guardian. They can also talk to friends, religious advisers, counselors, or other responsible adults. If one person does not believe them or isn't interested, victims should find another person to talk to.

Victims need to get any physical injuries treated. Even if the injuries seem small, it's okay to ask for help. Some injuries can affect a person throughout his or her life. The victim may need regular medical care.

Victims are often harmed mentally, too. They may get angry and become violent with others. They may believe that they deserved to be hurt. If they believe this, they may not try to stop the violence. Nobody deserves to be hurt.

Victims may feel afraid. They may fear that the people who hurt them will hurt them again. They may think they were to blame. All these feelings can be painful and frightening. A counselor can help victims deal with mental and emotional harm.

It's important for a person who has been a victim of violence to know that others care. Someone could listen if the person wants to talk, or invite him or her to do things. Knowing that there is someone who will listen and offer support can help the person recover.

List two things that a victim can do to help himself or herself recover from violence.

LESSON REVIEW

Review Concepts

1. **List** three examples of violence in a community.
2. **Identify** three ways you can prevent violence or fighting.
3. **Describe** how a victim of violence can get help.

Critical Thinking

4. **Evaluate** Some people say that violence in the media changes how people behave. Do you agree? Explain your answer.
5. LIFE SKILLS **Use Communication Skills** Suppose a student is angry about something that happened at school. How could he or she talk to his or her parents about this?

Learning LIFE SKILLS

CRITICAL THINKING

Resolve Conflicts

Problem A classmate blamed Emma for breaking a window that she didn't break. Now Emma wants to fight the other girl. How can she avoid violence?

Solution Resolving conflicts can help reduce the risk of violence. You can use conflict resolution skills to settle a disagreement.

Learn This Life Skill

Follow these steps to help you resolve conflicts.

1 Stay calm.

Emma is very angry with her classmate. How can she stay calm? She can take deep breaths and cool off before she does anything. What else could she do to keep control of her anger?

2 Talk about the conflict.

Once Emma calms down, she can talk to her classmate about the conflict. She can use I-messages to explain how she feels. She might say, "I feel angry when you blame me for breaking the window when I didn't do it." How could she use active listening skills?

3 List possible ways to settle the conflict.

Emma and her classmate could apologize to each other. What are some other ways they might settle the conflict?

4 Agree on a way to settle the conflict. You may need to ask a responsible adult for help.

Use the *Guidelines for Making Responsible Decisions*™ to evaluate the possible ways to settle the conflict. What would be the most responsible choice for Emma? When might she need to ask an adult for help?

Practice This Life Skill

Suppose that someone drops one of your books in a puddle of water on purpose. Work with a partner to role-play ways to resolve the conflict without violence. Use all four steps in your role-play.

LESSON 6

Steering Clear of Gangs

You will learn . . .

- what characteristics define a gang.
- ways to avoid gangs and weapons.

Vocabulary

- gang, *C75*

Do you have a group of friends you enjoy spending time with? Feeling like you are part of a group is important. The actions of some groups are not healthful, though. You can learn how to stay away from groups that can cause problems.

What Is a Gang?

A **gang** is a group of people often involved in dangerous and illegal acts. Gang members can be young people or adults. Both boys and girls join gangs.

People join gangs for many reasons. They might join a gang because they are having problems at home, at school, or with friends. They may want to be part of a group.

Gang members may wear specific colors, use hand signals, and have tattoos. They use these as symbols of the gang. They may draw graffiti. Graffiti is also used as a symbol of the gang. Gang members may steal money. They may use the money to buy illegal drugs. They may use weapons to fight with or kill members of other gangs. They may even kill people who are not in any gang.

Why do people join gangs?

Character

Include Others

Fairness You read about cliques in Chapter 2. In many ways a clique is like a gang. A clique will often have a certain look or behave in a certain way. People in a clique may reject others who look or act differently. Role-play a situation in which you include someone instead of excluding them.

Work toward including others, not excluding them. That way gangs and cliques won't form.

Design Pics/Ron Nickel

How to Stay Out of Gangs

ACTIVITY

Music LINK

Sing Against Gangs

Some popular music makes gangs sound cool. The songs often have lyrics about weapons and violence. With a partner, write song lyrics against gangs. Include lyrics that suggest ways to make friends outside of gangs. Perform your song for your class.

Staying away from gangs can help keep you safe. Gang members may expect other members to behave in illegal and dangerous ways. These behaviors can harm or even kill gang members and others. Even if a gang member is not hurt, he or she can go to jail. This can make it hard to finish school or get a job. The person may let down his or her family. It's important to follow family guidelines and choose legal, safe activities.

How can you steer clear of gangs? Here are some tips.

- **Stay away** from gang members. That way they can't pressure you to join the gang.
- **Stay away** from places where gang members hang out. You will be less likely to get caught up in violence.
- **Say "no" and walk away** if someone pressures you to join a gang.
- **Spend time with your family.** Your family can support you when you feel pressured.
- **Attend school** and school activities. These are healthful ways to spend time.
- **Find hobbies** that you like to do. Join a club or sports team. You can make new friends who have the same interests you do.

▲ Joining school activities can help you avoid gangs.

What You Should Know About Weapons

Some people carry weapons because they feel afraid. They think they can use the weapons to defend themselves. But people who carry weapons are more likely to be hurt by violence. They may not choose other ways to resolve conflicts.

It's better not to spend time with people who carry weapons. Try to get away from a person with a weapon. Don't argue with him or her.

Be careful around weapons. Don't touch a gun without permission. Never touch a gun you find outside or in someone else's home. It might be loaded. You could hurt someone by mistake. Tell an adult if you find a weapon.

Why is it wise to avoid weapons?

ACTIVITY
LIFE SKILLS
CRITICAL THINKING

Use Resistance Skills

In a small group, design a poster that shows how to use resistance skills to avoid gangs. Include these steps in your poster.

1. **Look at the person. Say "no" in a firm voice.** Show other students saying "no" to gangs in your ads.
2. **Give reasons for saying "no."** Write catchy slogans that give reasons not to join a gang.
3. **Match your behavior to your words.** Show a student walking away from a gang.
4. **Ask an adult for help if you need it.** In your poster, remind students to talk to parents, guardians, or teachers if they need help.

LESSON REVIEW

Review Concepts

1. **List** two examples each of who gang members are and what gang members do.
2. **Identify** three things you can do to avoid gangs.

Critical Thinking

3. **Convince** Suppose that a friend wants to get a weapon. Write a paragraph to persuade your friend not to do this.
4. **Infer** Explain how not joining a gang can help you lead a healthful life.
5. LIFE SKILLS **Use Resistance Skills** Suppose a person who is in a gang wants you to join. How can you use resistance skills to get out of the situation?

CHAPTER 6 REVIEW

Use Vocabulary

emergency, *C55*
first aid, *C61*
frostbite, *C52*
heatstroke, *C52*
safety rules, *C43*
smoke detector, *C44*
tetanus, *C61*
universal precautions, *p. C61*

Choose the correct term from the list to complete each sentence.

1. Guidelines that help prevent injury are _?_.
2. A person who has a sudden illness or injury may need quick and temporary care called _?_.
3. Help is needed quickly in a(n) _?_.
4. The steps taken to avoid having contact with pathogens in body fluid are called _?_.
5. A person who gets too hot may suffer from _?_.
6. Your house should have a(n) _?_ on each floor.
7. An injury caused by exposure to extreme cold is _?_.
8. A disease that attacks the nervous system and is caused by bacteria that enter a wound is _?_.

Review Concepts

Answer each question in complete sentences.

9. What are the signs of frostbite? What should you do if you see these signs?
10. Name five items that should be placed in a disaster kit.
11. What is first aid for a bee sting?
12. List three ways to prevent violence.
13. Name four emergency situations when you should call for help.
14. Describe the steps to follow if there is a fire at home.

Reading Comprehension

Answer each question in complete sentences.

Victims are often harmed mentally, too. They may get angry and become violent with others. They may believe that they deserved to be hurt. If they believe this, they may not try to stop the violence. Nobody deserves to be hurt.

15. How might a victim of violence feel?
16. Why would a victim not try to stop the violence?
17. Why is it important for a victim to get help?

Critical Thinking/Problem Solving

Answer each question in complete sentences.

Analyze Concepts

18. How can you keep from fighting if you are bullied?

19. Describe an emergency situation that might happen at school. Describe what you should do in the emergency.

20. A friend sprains his ankle while you are playing basketball. What should you do?

21. Name three types of violence and give an example of each.

22. Suppose you want to go for a walk, but the neighborhood doesn't have sidewalks. What could you do to be safe on your walk?

Practice Life Skills

23. **Resolve Conflicts** Your classmate calls you stupid for missing a word in a spelling bee. You get angry. You even feel like hitting him for calling you a name. How can you resolve the conflict without violence?

24. **Make Responsible Decisions** Your best friend is spending time with students who get into trouble at school. She tells you how cool her new friends are. She brags about how much fun they have together. She wants you to join the group. What should you do? Use the *Guidelines for Making Responsible Decisions*™ to help you decide.

Analyze Graphics

Use the chart to answer the questions.

Safety Precautions

	At Home	Outdoors
Thunderstorm	Don't use electrical items. Don't talk on the phone.	Get inside. Avoid tall trees.
Tornado	Stay away from windows. Take shelter in the basement if possible.	Get out of a car. Lie down in a low-lying area if you can't get inside.
Hurricane	Board up windows. Be ready to evacuate.	Get inside as soon as possible. Avoid objects that could blow onto you.
Earthquake	Drop to the floor. Take cover under a table or desk.	Get away from buildings and electrical wires. Cover your head to protect it from falling objects.

25. When should you board up windows?

26. When should you be careful about electrical items?

27. Explain the safety rules shown for being indoors during a tornado.

ACTIVITIES AND PROJECTS

Effective Communication

Make a Video

Videotape a friend demonstrating the safety rules for walking along a street. Narrate the video as you show it to your class.

Self-Directed Learning

Design a Booklet

Make a booklet telling others how to care for a body part. Choose your teeth, eyes, or ears. Find out more about how to take care of the body part you chose.

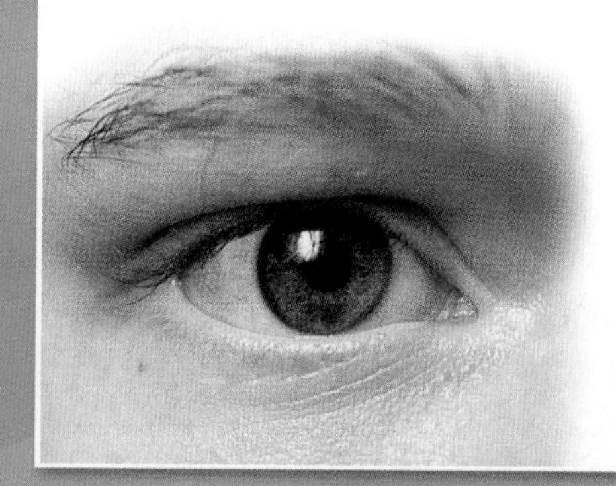

Critical Thinking and Problem Solving

Graph Activities

Survey your classmates on their favorite physical activity. Make a graph showing how many students chose each activity. Label the graph with captions telling which kinds of health fitness each activity improves.

Responsible Citizenship

Invite a Guest Speaker

With your teacher's permission, contact your local fire department. (Call the regular office number, not 9-1-1.) Invite a firefighter to your classroom to discuss fire safety.

UNIT D

Drugs and Disease Prevention

CHAPTER 7

Alcohol, Tobacco, and Other Drugs, *D2*

CHAPTER 8

Communicable and Chronic Diseases, *D42*

CHAPTER 7

Alcohol, Tobacco, and Other Drugs

Lesson 1 • Drugs and Your Health. *D4*

Lesson 2 • Alcohol and Health. *D10*

Lesson 3 • Tobacco and Health. *D16*

Lesson 4 • Other Drugs to Avoid *D24*

Lesson 5 • When Someone Abuses Drugs . . . *D30*

Lesson 6 • Resisting Pressure. *D36*

What Do You Know?

What do you know about the health effects of smoking? Write **T** for each true statement and **F** for each false statement.

___?___ Nicotine and tar from tobacco products can stain your hands and teeth.

___?___ Smoking can cause such diseases as lung cancer and emphysema.

___?___ Heart attacks and strokes happen more often to people who smoke than to people who don't.

___?___ Other people's tobacco smoke can increase your risk of cancer.

___?___ Smoking can make it difficult to breathe.

Believe it or not, all five statements are true. Smoking is a habit that can harm your health in many ways. Learn more about this topic by reading **Alcohol, Tobacco, and Other Drugs**.

LESSON 1

Drugs and Your Health

You will learn . . .

- how medicines are used to promote health.
- safety rules for using medicine.
- the signs of drug misuse and abuse.
- four steps to drug dependence.

Vocabulary

- **drug**, *D5*
- **medicine**, *D5*
- **over-the-counter (OTC) drug**, *D5*
- **prescription drug**, *D5*
- **side effect**, *D7*
- **drug misuse**, *D8*
- **illegal drug**, *D8*
- **drug abuse**, *D8*
- **addiction**, *D8*

You've probably taken medicine when you were ill. Medicines can improve your health if you use them properly. They can harm your health if you don't use them in safe ways.

Drugs Used as Medicine

A **drug** is a substance that changes how the mind or body works. A **medicine** is a drug used to prevent, treat, or cure a health condition.

Your parents or guardian may buy medicine without a doctor's order. **Over-the-counter (OTC) drugs** are medicines that can be bought without a doctor's order. A medicine that can be obtained only with a doctor's order is a **prescription drug**.

There are different kinds of medicines. *Antibiotics* help your body fight off some infections. *Vaccines* keep you from getting some illnesses. Other medicines reduce pain or swelling.

Many drugs have brand names. Companies think the brand names will make people want to buy their drugs. Some drugs are generic drugs. They have no brand names but have the same main ingredients as brand-name drugs. They are often cheaper than drugs with brand names. A generic drug and a brand-name drug may do the same thing.

What do antibiotics do?

ACTIVITY

Consumer Wise

Design a Brochure

Make a brochure that compares a brand-name medicine and a generic medicine. With your parents' or guardian's permission, visit a grocery store or pharmacy. Choose a brand-name medicine and a generic medicine that matches it. Read the labels on both medicines. Write down the ingredients, the dose, what the medicine does, and the cost. Use your information to make a brochure comparing and contrasting the two medicines. Share your brochure with your class.

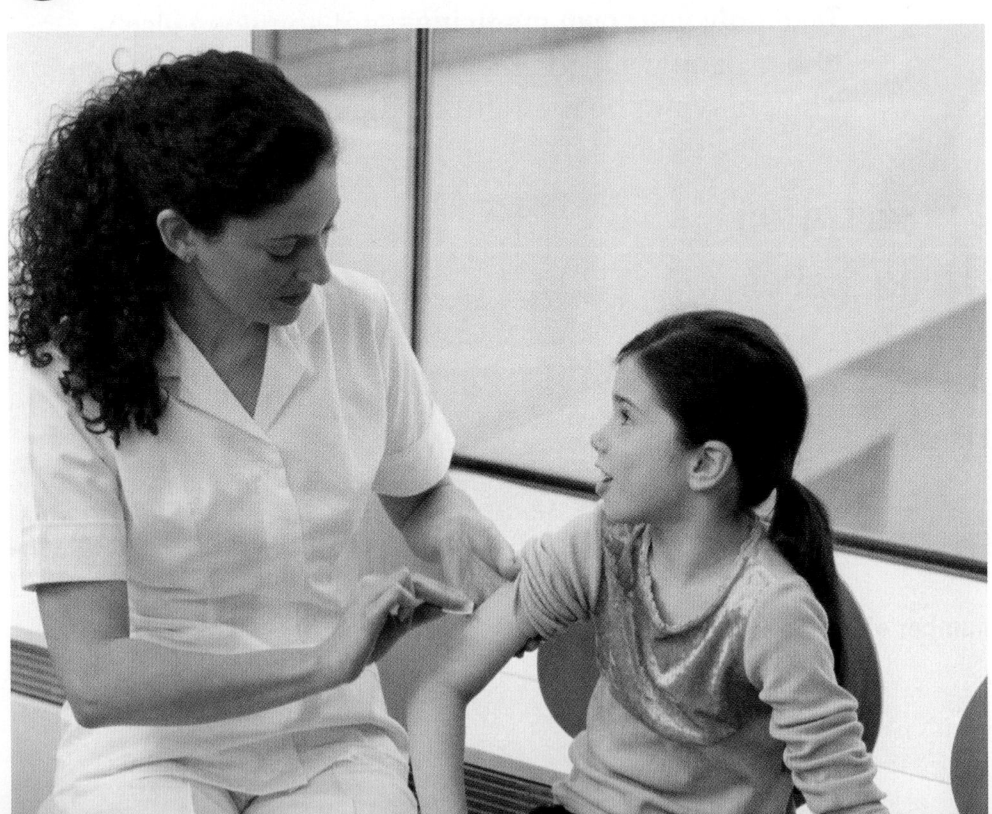

Vaccines can protect you against diseases, including chicken pox, mumps, measles, and tetanus.

Safety Rules for Medicine

Companies that sell OTC medicines must label them with safe doses. If the safe dose is different for children, they must say so on the label.

Medicines can help your health. But they can harm you if they aren't used safely. Here are some important safety rules for using medicines.

- **Take medicine only from your parents or guardian** or another adult they approve of. Your parent or guardian may give permission to someone else to give you medicine if he or she is unable to. Your parent or guardian will tell you who may give you medicine.
- **Follow the instructions** that come with your medicine. Take the right dose. A *dose* is the amount of medicine you should take. Your doctor may tell you the dose. An OTC package may list a dose. The label also may tell you how to take the medicine. You may need to take it with food. You may need to drink lots of water with the medicine.
- **Tell your doctor and pharmacist** about any medicines you take. Sometimes drugs can change the action of other drugs when taken together. These changes can make the drugs not work properly. They also can harm you.
- **Take only your own medicines**, not someone else's. Other people's medicines may be too strong for you. They may not be the right medicine for you.

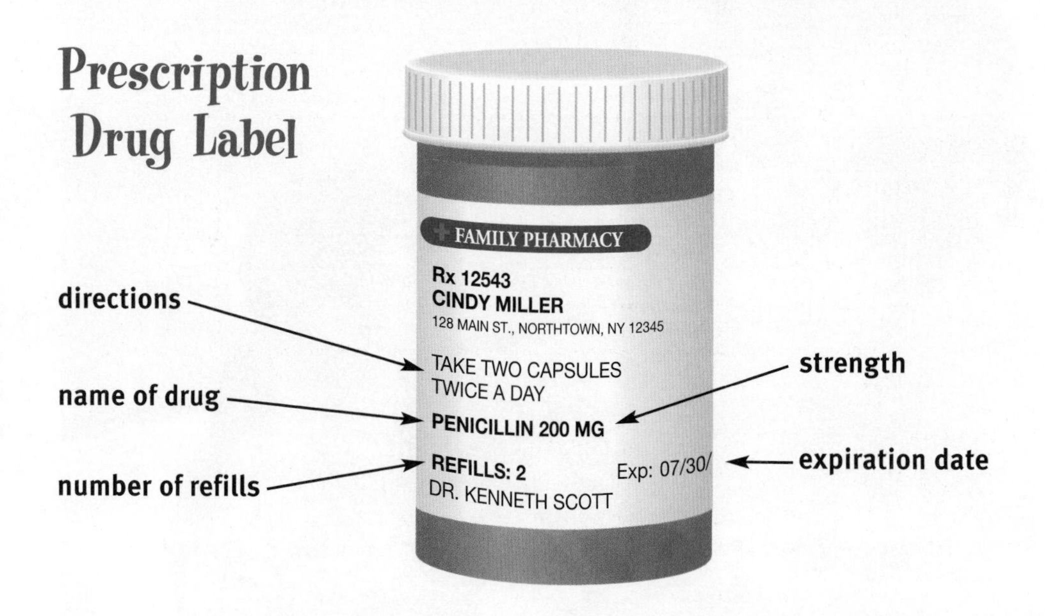

- **Stop taking the medicine** and tell your parents or guardian right away if you have any side effects. **Side effects** are unwanted reactions to a drug. A drug might make you sleepy. Another drug might upset your stomach.
- **Keep medicines in their containers.** The containers have instructions for using the medicines safely. OTC medicines usually have a date stamped on the package. Don't use the medicine after that date.
- **Check the seal** on packages of OTC medicine. The seal protects the medicine. Do not buy OTC medicine if the seal is broken.

What are side effects?

OTC Medicine Label

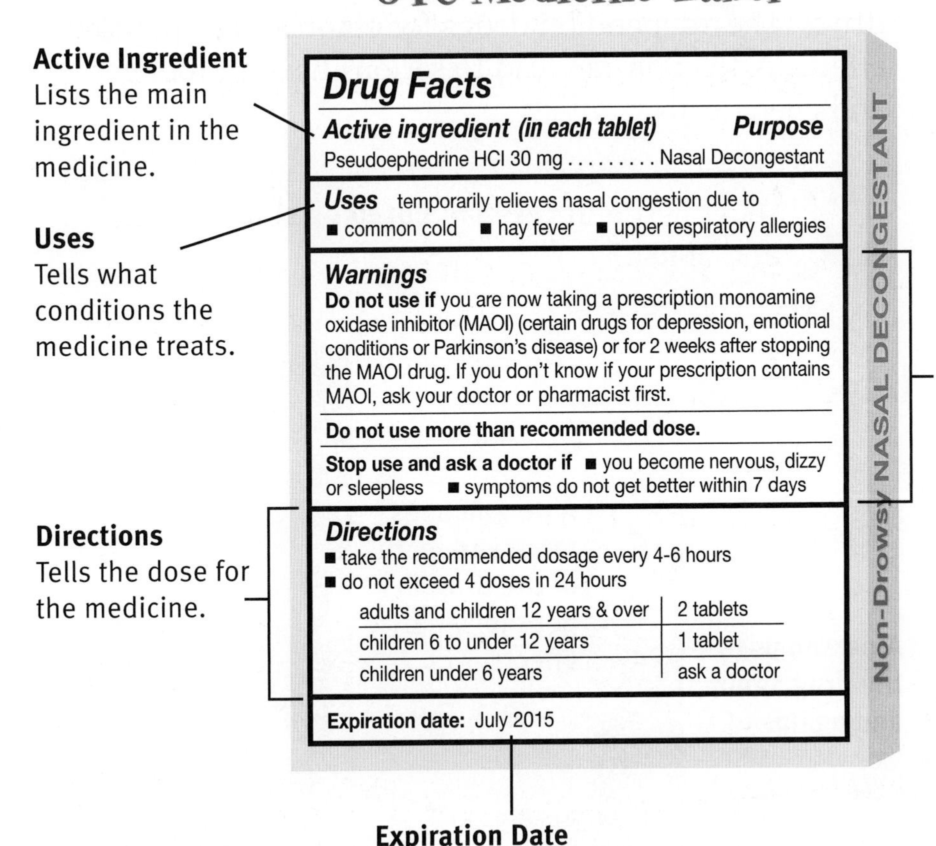

ACTIVITY

On Your Own

FOR SCHOOL OR HOME

Find Expiration Dates

With your parents or guardian, read the labels on OTC medicines in your home. Make a chart showing the expiration date for each medicine. Post the chart near where you store the medicines. Check it often to see whether you need to throw out any medicines. Every time you check it, mark on your chart which medicines are still good.

Drug Misuse and Abuse

Write About It!

Write a Commercial Sniffing glue, chewing tobacco, and taking steroids to build muscles are all examples of drug abuse. With a group, choose one form of drug abuse. Write a commercial telling the harmful effects of abusing the drug you chose. Perform your commercial for the class.

You have an ear infection. A doctor prescribed medicine for you. You are supposed to take two pills each day. You take four pills by mistake. This is drug misuse. **Drug misuse** is the accidental unsafe use of a medicine. If you misuse a drug, tell your parents or guardian right away. Your parent or guardian can decide what to do. He or she may call a doctor.

Some people use legal drugs in unsafe ways on purpose. They like how the drugs make them feel. Other people use illegal drugs. An **illegal drug** is a drug that is against the law to have, use, buy, or sell. The use of an illegal drug or the harmful use of a legal drug is called **drug abuse**. Abusing drugs can harm your mind and your body.

Drug abuse can lead to an addiction. An **addiction** is a strong desire to do something even though it is harmful. People who have a drug addiction are dependent on drugs. They depend on drugs for emotional or physical effects. Addiction can lead to serious health problems and even death.

What is the difference between drug misuse and abuse?

Four Steps to Drug Dependence

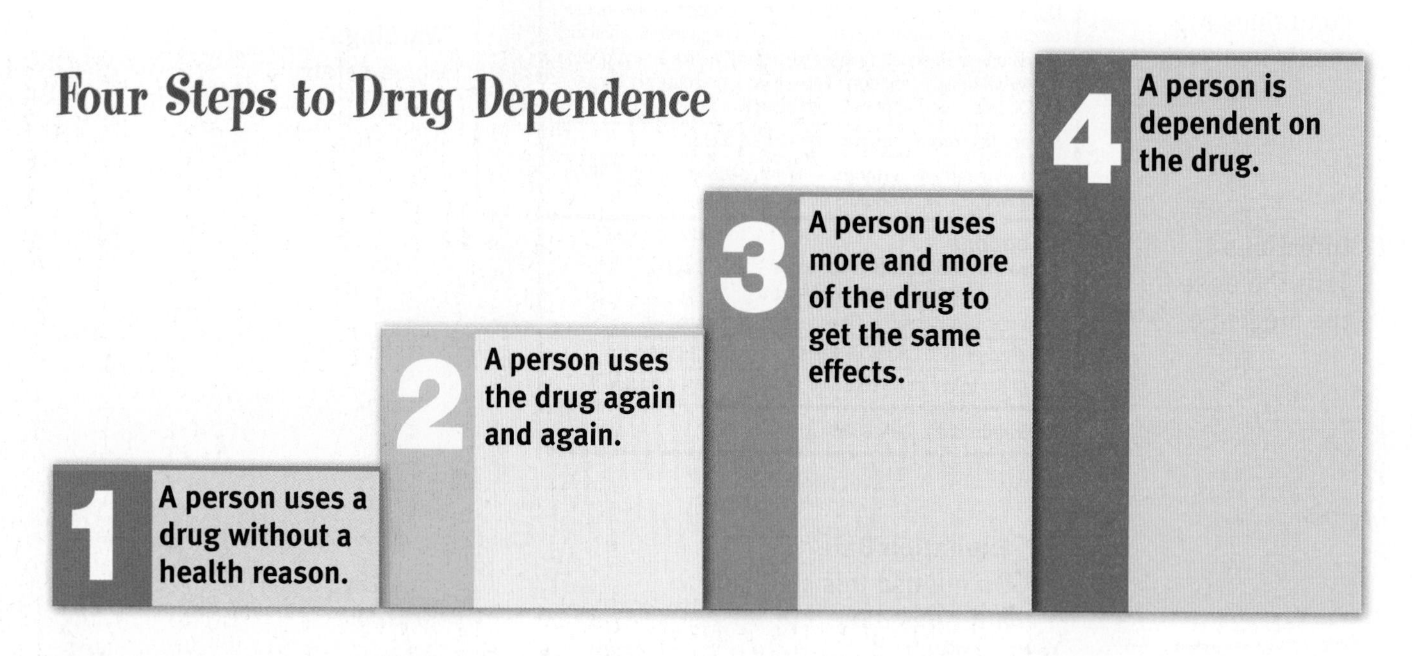

ACTIVITY

LIFE SKILLS CRITICAL THINKING

Make Responsible Decisions

Ron is having trouble paying attention in school. Ashley tells him that a prescription medicine she takes helps her concentrate. She offers Ron her pills. Should he take them? Make a diagram that shows how Ron could use the *Guidelines for Making Responsible Decisions*™ to decide.

1. **Identify your choices. Check them with your parent or trusted adult.** Ron could take Ashley's medicine. He could refuse to take it. Draw one circle for each choice. Write the choice inside the circle.
2. **Evaluate each choice. Use the *Guidelines for Making Responsible Decisions*™.** Ask yourself each question. Put a star in the circle for each "yes" answer. Put an X in the circle for each "no" answer.
3. **Identify the responsible decision. Check this choice out with your parent or trusted adult.** Which choice has six stars? That is the most responsible decision.
4. **Evaluate your decision.** Below each circle, write why the choice was or was not responsible. What are the benefits of a responsible decision?

***Guidelines for Making Responsible Decisions*™**

- **Is it healthful?**
- **Is it safe?**
- **Does it follow rules and laws?**
- **Does it show respect for others and myself?**
- **Does it follow family guidelines?**
- **Does it show good character?**

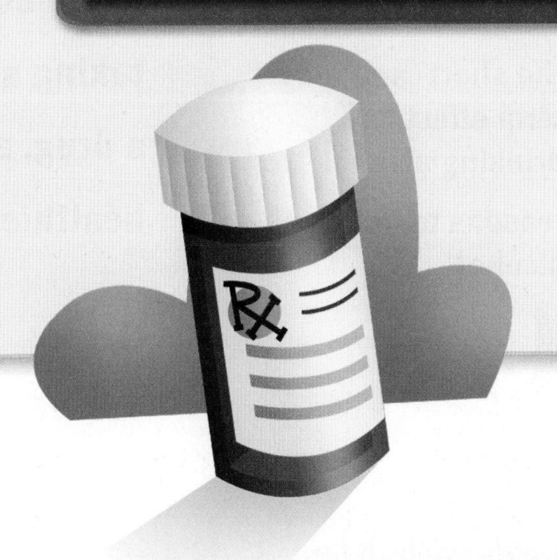

LESSON REVIEW

Review Concepts

1. **Describe** the difference between prescription and over-the-counter drugs.
2. **Explain** what to do if you experience a side effect of a drug.
3. **List** the four steps that lead to drug dependence.

Critical Thinking

4. **Infer** Why is it difficult for a person with an addiction to stop using a drug?
5. LIFE SKILLS **Make Responsible Decisions** You and your sister are taking different prescription medicines. You accidentally take one of her pills instead of one of yours. What should you do?

LESSON 2

Alcohol and Health

You will learn . . .

- what alcohol is and its effects on the body.
- the short- and long-term effects of drinking alcohol.
- reasons to not drink alcohol.

Vocabulary

- **alcohol**, *D11*
- **depressant**, *D11*
- **blood alcohol concentration (BAC)**, *D11*
- **intoxication**, *D11*
- **alcoholism**, *D12*
- **minor**, *D12*

When someone drinks beer, he or she is actually taking a drug. Beer contains alcohol. Alcohol is a drug. Staying alcohol free can help you live a healthier life.

What Is Alcohol?

Alcohol is a depressant drug found in some beverages. A **depressant** is a drug that slows down body functions. Alcohol slows down a person's brain and body. At first it may make a person feel relaxed.

Why do people drink alcohol? They may want to fit in with a group. They may want to feel older. They may think it will help them avoid their problems. None of these are healthful reasons.

Alcohol and the Body

When a person drinks, some alcohol enters the blood right away. Alcohol enters the blood more quickly if a person hasn't eaten in a while.

The liver is responsible for breaking down alcohol so that the body gets rid of it. The liver can only handle a limited amount of alcohol at once, though. For most adults, the liver can handle about as much alcohol each hour as is in one beer. If a person drinks more alcohol than the liver can handle, the rest of the alcohol builds up in the blood.

The amount of alcohol in a person's blood is called the **blood alcohol concentration (BAC)**. A BAC of 0.01 means that one ten-thousandth of the person's blood is alcohol. The BAC depends on a person's weight and how much alcohol the person drank. It also depends on how much time the person spent drinking.

A high BAC can lead to intoxication. **Intoxication** is the state of being drunk. Intoxication makes it difficult to think clearly. It slows down a person's reaction time. It decreases a person's coordination. People who are intoxicated are more likely to have car accidents. In most states, it is illegal for a person with a BAC of 0.08 or higher to drive a car.

▲ **One 12-ounce beer has about the same amount of alcohol as a 4-ounce glass of wine or about 1 ounce of liquor.**

What kind of drug is alcohol?

Effects of Alcohol

Write About It!

Write a Persuasive Song Write a song or poem persuading other young people not to use alcohol. Include at least two short-term and two long-term effects as reasons to avoid alcohol. Use reasons that you think will convince other people your age not to drink alcohol. Share your song or poem with your class.

Drinking alcohol can harm your physical, mental and emotional, and family and social health. One possible effect of drinking alcohol is alcoholism. **Alcoholism** is a disease in which a person is addicted to alcohol. The signs of the disease aren't the same in everyone who has it. Some people with alcoholism drink often. Some do not. A person who has family members with alcoholism has a higher risk for the disease. People who start drinking at a young age also increase their risk.

It is against the law for minors to drink alcohol. A **minor** is a person under the legal age for an action such as drinking alcohol. You are a minor.

Short-term Effects of Alcohol

Alcohol has both short- and long-term effects. Here are some ways it can affect you in the short term.

Difficulty Thinking Clearly Drinking alcohol makes it harder to make responsible decisions. It also harms the memory. It makes it harder to do well in school.

Poorer Coordination and Slower Reaction Time You won't do as well at sports if you drink alcohol. People who drink alcohol are more likely to have car accidents.

Stronger Emotions Alcohol makes emotions seem more intense. You might get extra angry or extra sad.

Breaking Family Guidelines Your parents or guardian do not want you to drink alcohol.

Long-term Effects of Alcohol

Drinking alcohol can also have long-term effects. People who start drinking at a young age can have health problems later in life. Some of these problems can shorten a person's life.

Liver Disease The liver has to work very hard to remove alcohol from the blood. Over time the extra work can cause damage to the liver and diseases such as cirrhosis (suh•ROH•suhs) and alcoholic hepatitis (he•puh•TIGH•tuhs).

Heart Disease and Cancer Heavy drinking can also increase the risk of heart disease over time. The heart does not pump blood as well. Drinking alcohol also increases the risk of cancer of the mouth, throat, esophagus, liver, colon, and breast.

Damage to Organs Drinking alcohol can harm the pancreas. This important gland can become *inflamed,* or swollen. Heavy alcohol use can also harm the kidneys, bones, and muscles. It also causes memory loss.

Birth Defects Women who drink while they are pregnant risk harming their babies. Birth defects caused by a mother using alcohol are called *fetal alcohol syndrome.*

Social Problems Alcohol makes it difficult to make responsible decisions. People who abuse alcohol are more likely to be involved in violence. This can lead to injuries and even death. People who drink heavily may also have trouble using social skills. They have trouble expressing their emotions in healthful ways.

ACTIVITY

Science LINK

Draw a Body Map

On a poster board, draw a map of the human body. Use different colors to show the organs that can be harmed by drinking alcohol. Include both short-term and long-term effects. Label each organ with how it can be harmed by alcohol.

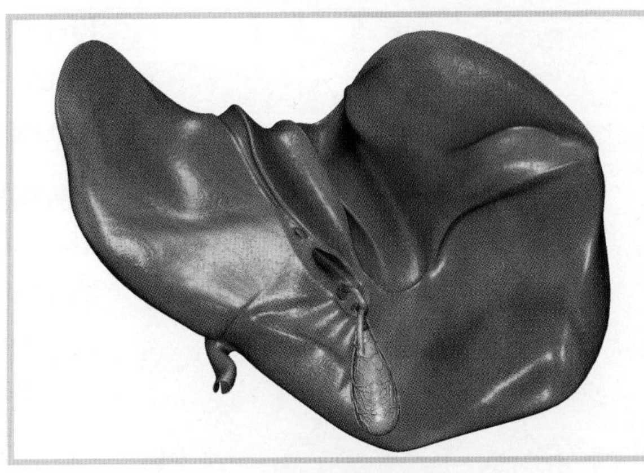

healthy liver

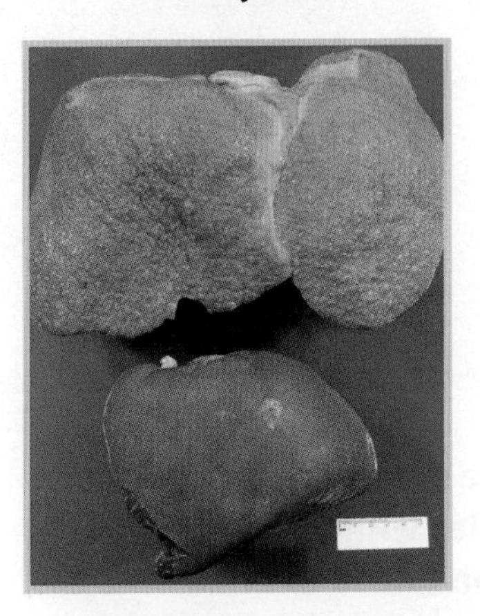

liver with cirrhosis

▲ **Alcoholic cirrhosis is a scarring of the liver. Cirrhosis is not reversible. A person with this disease may need a liver transplant to survive.**

Reasons Not to Drink Alcohol

How can you say "no" if someone pressures you to drink alcohol? It may help you say "no" if you have reasons to give for not drinking alcohol.

I want to obey laws. It's against the law for people your age to buy or drink alcohol. You must be at least 21 years old to buy or drink alcohol legally.

I want to make responsible decisions. Alcohol can make you think less clearly. You might say or do something that you would regret later.

▲ Choose fun and healthful activities with your friends. You don't need to drink alcohol to have fun.

I want my senses to function at their best. Drinking alcohol can change how your senses work. Your vision can become blurred. This makes it hard to see well and to judge distances.

I want to keep from having accidents. Drinking alcohol slows down your reaction time. Suppose that you are riding your bicycle and someone steps in front of you. It would take you longer to brake or change directions if you had been drinking alcohol. You'd be more likely to hit the person.

I want to have healthful relationships. Drinking alcohol changes how you respond to other people. People who depend on alcohol don't use their social skills as much. They can't think as clearly, so they don't make responsible choices about their relationships.

I want to get along well with my family. Your parents or guardians don't want you to drink. You will go against family guidelines if you drink alcohol. Your family will be angry with you if you get caught.

I want to keep from being depressed. If you feel sad, alcohol can make you feel more sad. It may keep you from dealing with your problems in healthful ways.

I want to stay away from fights. Drinking alcohol affects how your brain deals with emotion. You may not manage anger well. You are less likely to stop yourself from getting into fights.

I want to protect myself from diseases such as heart disease, cancer, and liver disease. Drinking alcohol can harm body cells and organs. It can increase your risk of some diseases.

I do not want to have alcoholism. Alcoholism can harm your life and the lives of people around you. You might be more likely to develop alcoholism if someone in your family has the disease. You can prevent alcoholism by staying alcohol free.

How does alcohol use lead to more accidents?

ACTIVITY

LIFE SKILLS

CRITICAL THINKING

Set Health Goals

Set a health goal not to drink alcohol. Make a Health Behavior Contract.

1. **Write the health goal you want to set:** I will not drink alcohol. Write your health goal in your health behavior contract.
2. **Explain how your goal might affect your health.** Write both a short-term effect and a long-term effect of alcohol use that you want to avoid.
3. **Describe a plan you will follow to reach your goal. Keep track of your progress.** Choose three reasons from the list on pages D14 and D15. Write them on your Health Behavior Contract. If anyone asks you why you want to be alcohol free, give one of your reasons. Put a mark next to it when you use it.
4. **Evaluate how your plan worked.** Which reasons did you use most often? Write a sentence or two explaining why you chose those reasons.

ALCOHOL FREE KID!

LESSON REVIEW

Review Concepts

1. **Define** alcohol.
2. **List** three short-term and three long-term effects of drinking alcohol.
3. **Identify** five reasons to not drink alcohol.

Critical Thinking

4. **Synthesize** Explain how drinking alcohol can increase the risk of injuries and disease.
5. LIFE SKILLS **Set Health Goals** You have set a health goal not to drink alcohol. What are steps you can take toward your goal?

LESSON 3

Tobacco and Health

You will learn . . .

- the harmful effects of toxins found in tobacco smoke.
- the short- and long-term effects of tobacco use.
- how to quit tobacco use.
- how secondhand smoke can be harmful to health.

Vocabulary

- **nicotine**, *D17*
- **tar**, *D17*
- **carbon monoxide**, *D17*
- **smokeless tobacco**, *D17*
- **secondhand smoke**, *D20*

Each year about 400,000 people die from smoking-related diseases in the United States. You can reduce your risk of many diseases by not smoking.

What Is Tobacco?

Tobacco is a plant that contains nicotine. Its leaves are dried and then made into cigarettes, cigars, smokeless tobacco, and other products. People smoke, chew, or inhale these tobacco products. Tobacco is legal for adults. It is illegal for people your age. It is harmful for everyone.

Tobacco smoke contains more than 4,000 chemicals. Many of these chemicals are *toxins,* or poisons. Three of these toxins are nicotine, tar, and carbon monoxide.

- **Nicotine** (NI•kuh•teen) is a stimulant drug found in tobacco. Stimulant drugs speed up body functions. Nicotine is very addictive.
- **Tar** is the gummy substance found in tobacco smoke. Tar can kill lung cells.
- **Carbon monoxide** (KAR•buhn muh•NAHK•sighd) is a poisonous gas found in tobacco smoke. This gas makes it hard for blood to carry oxygen.

Nicotine and tar are also in smokeless tobacco. **Smokeless tobacco** is tobacco that is chewed or placed between the cheek and gums. High levels of nicotine enter the body through the gums.

How does tar affect your lungs?

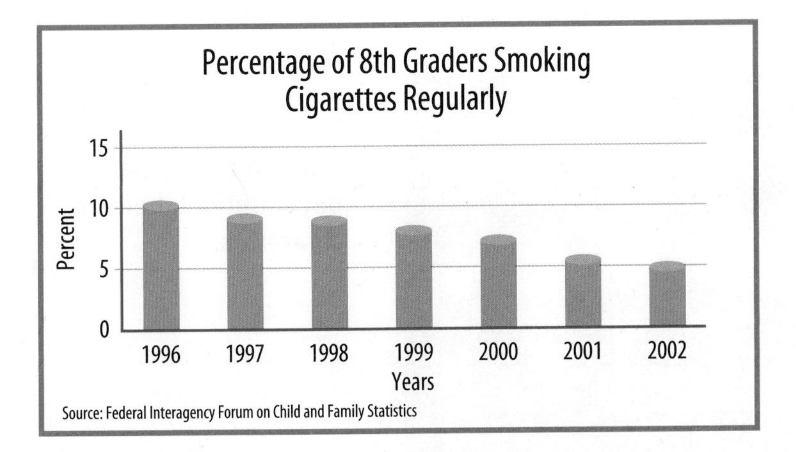

▲ **According to the graph, how has the number of young people who smoke changed since 1996?**

ACTIVITY

Art LINK

Make an Anti-Cigarette Ad

Design your own ad against cigarettes. In the ad, tell what harmful chemicals are in cigarettes and the health benefits of not smoking. Share your ad with the class.

Harmful Effects of Tobacco

Explore the Effects of Smoking

Take one regular drinking straw and put it in your mouth. Breathe through the straw for 30 seconds. Next, take a smaller straw, such as a coffee stirrer, and breathe through it for 30 seconds. Did you have difficulty breathing through either straw? Breathing through the larger straw shows how much harder the lungs have to work when a person has smoked for a short time. Breathing through the smaller straw shows how difficult it can be to breathe when a person has smoked for years.

Some people might start smoking because they want to fit in, or because they like the way it makes them feel. But tobacco harms the body in many ways.

Short-term Effects of Tobacco

Tobacco affects the body very quickly. Here are some short-term effects of using tobacco.

Addiction The nicotine in tobacco is very addictive.

Breathing Problems The blood of someone who uses tobacco can't carry as much oxygen. The person gets short of breath more easily. This makes it harder to be physically active or participate in sports.

Stains and Smells Nicotine and tar can stain hands, teeth, and clothes. Smoking also makes breath, clothes, and hair smell bad.

Cost Using tobacco is expensive. A person who has to keep buying tobacco products doesn't have that money to spend on other things.

Fires and Burns Smoking increases the risk of fires and burns. People who smoke may fall asleep while holding a cigarette. They may drop cigarettes on the ground. These can start fires. The lit end of a cigarette can burn the skin.

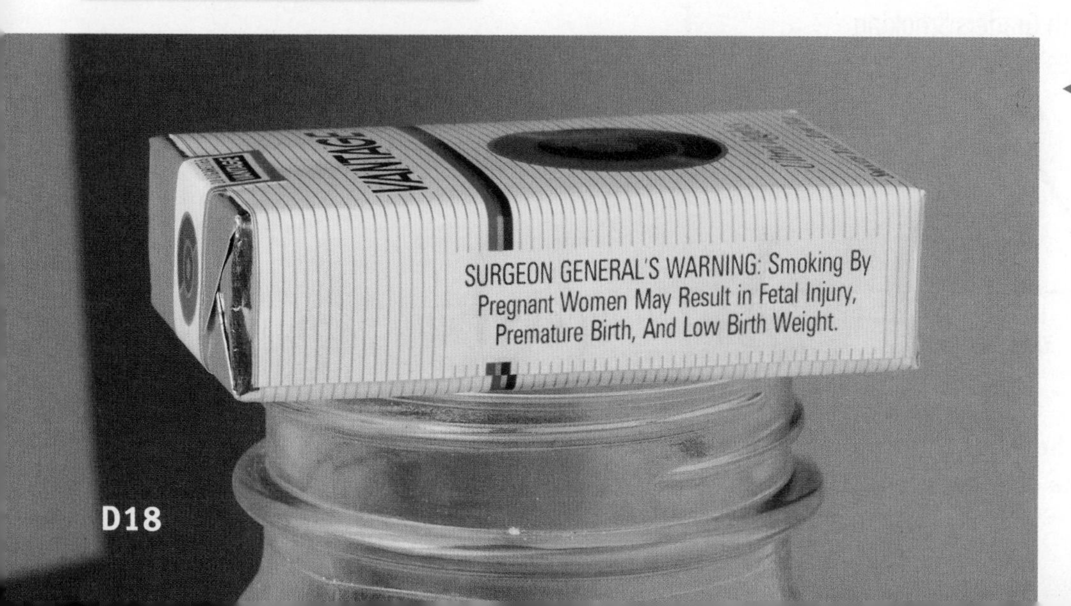

The government requires tobacco companies to put warning labels on cigarettes.

Jack Holtel/McGraw-Hill Education

Long-term Effects

Over time people who smoke are 1.5 times as likely to die early from disease as nonsmokers. They are more at risk for many health problems.

Heart Disease Carbon monoxide from tobacco makes the heart work harder. Nicotine makes the heart beat faster. This increases the risk of heart disease.

Lung Disease Tobacco smoking causes more than three-quarters of all cases of lung cancer. People who use smokeless tobacco increase their risk of mouth and throat cancer. Smoking also causes most cases of *emphysema* (emp•fuh•ZEE•muh), a disease in which the alveoli in the lungs are destroyed. Smoking tobacco also can keep a young person's lungs from growing. People who smoke often have more sore throats and coughs than nonsmokers.

Stomach Ulcers Ulcers are open sores. People who use tobacco are more likely to develop stomach ulcers.

Other Effects People who smoke are more likely than nonsmokers to have skin with wrinkles. Their gums are less healthy. Their teeth may fall out. Tobacco dulls their senses of taste and smell. Pregnant women who smoke are more likely to have babies with health problems.

Breaking Laws It is illegal for people under age 18 to buy tobacco.

How does smoking affect your teeth?

Calculate Tobacco Costs

If a pack of cigarettes costs about $3.50, and a smoker uses 2 packs a day, how much will this cost per week? If a CD costs $10, about how many CDs could the person buy for the same amount of money?

ACTIVITY

Music LINK

Write a No Smoking Song

With a partner, write a song encouraging people not to smoke in public places. Use your song to explain how secondhand smoke can harm others. Perform your song for the class.

How to Quit Tobacco Use

Nicotine is addictive. This makes it very hard to stop using tobacco. It's important to stop, though. A person who quits smoking starts to reduce his or her risk of disease in just one day.

There are programs to help tobacco users quit. The American Lung Association and American Cancer Society sponsor such programs. Hospitals and clinics often have programs. Medicines can help people who smoke deal with the physical effects of stopping smoking.

Secondhand Smoke

Have you been ever in a room or car with someone who was smoking? Did the smoke make your throat hurt? Did your eyes itch? Exhaled smoke and smoke from the burning end of a cigarette, cigar, or pipe is called **secondhand smoke**.

Secondhand smoke is a carcinogen (kar•SI•nuh•juhn). A *carcinogen* is something that can cause cancer. About 3,000 nonsmokers die every year from lung cancer caused by secondhand smoke. Secondhand smoke can cause heart disease. Young people who breathe secondhand smoke have a higher risk for lung problems. They get asthma more often. They also get more lung infections, such as pneumonia (new•MOHN•yuh) and bronchitis (brong•KIGH•tis).

Many places now have separate smoking rooms. This protects nonsmokers from secondhand smoke.

Years ago smoking was allowed almost everywhere. People did not know that secondhand smoke could cause health problems. Today smoking is banned in many office buildings and other places. Smoking is not allowed on most airplanes. Trains allow smoking only in certain cars. Hotels have rooms where smoking is not allowed. These changes have reduced your risk of diseases caused by secondhand smoke. Many communities have banned smoking in restaurants and other public places. These laws protect the people who work in these places. They also protect you and others who visit them.

What can you do when you are around secondhand smoke? You can politely ask people not to smoke around you. In places where smoking is not banned, look for a nonsmoking area. Many restaurants and other public places have nonsmoking sections. Sitting in these sections will help you avoid secondhand smoke.

Why is secondhand smoke harmful?

ACTIVITY

LIFE SKILLS

CRITICAL THINKING

Analyze What Influences Your Health

1. **Identify people and things that can influence your health.** Tobacco companies spend billions of dollars each year on ads. Find some tobacco-related ads in magazines. Cut them out. Tape or glue each ad to a large piece of paper.
2. **Evaluate how these people and things can affect your health.** Companies want you to think that smoking will improve your life. How do the ads you chose try to do this? Write a sentence below each ad explaining how it tries to influence you.
3. **Choose healthful influences.** Circle the warnings in the ads. How do they influence you in a healthful way?
4. **Protect yourself from harmful influences.** With your class, discuss the ads you found. What *don't* they tell you about tobacco use?

LESSON REVIEW

Review Concepts

1. **Describe** three toxins found in tobacco smoke and their effects on the body.
2. **Name** five diseases that are related to tobacco use.
3. **Explain** why it is healthful to avoid secondhand smoke.

Critical Thinking

4. **Evaluate** How do laws about tobacco use protect people your age?
5. LIFE SKILLS **Analyze What Influences Your Health** Why might tobacco companies not tell you all the facts about cigarettes in their ads?

Learning LIFE SKILLS

CRITICAL THINKING

Be a Health Advocate

Problem Jon, Elena, and Rachel read about the effects of secondhand smoke. They want to protect the health of the students in their school. What can they do?

Solution Being a health advocate means sharing health information with other people. Jon, Elena, and Rachel can be health advocates in their school. You also can be a health advocate.

Learn This Life Skill

Follow these steps to help you be a health advocate.

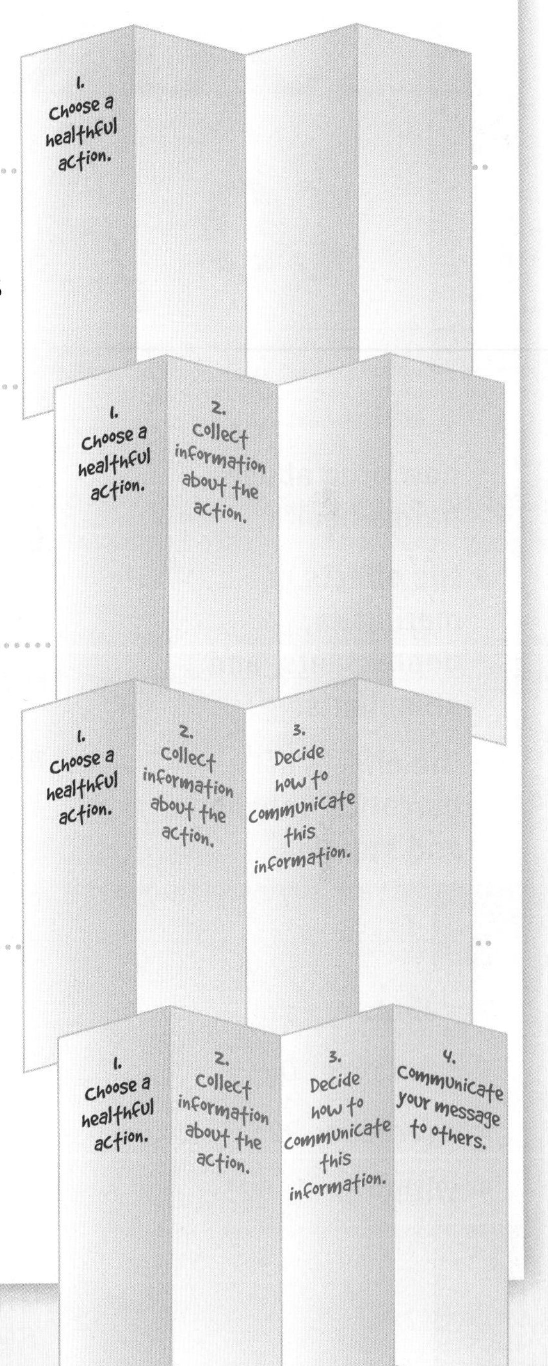

1 Choose a healthful action to communicate.

Jon, Elena, and Rachel want to protect themselves and others from secondhand smoke.

2 Collect information about the action.

Jon, Elena, and Rachel can list ways secondhand smoke harms health. Then they can list ways to avoid it. Make a list of your own.

3 Decide how to communicate this information.

Jon, Elena, and Rachel made posters to display in the school. They put facts from the list on the posters. What are some other ways they could communicate these facts?

4 Communicate your message to others.

Jon, Elena, and Rachel put their posters around the school where students would see them. What could you do?

Practice This Life Skill

With a group, brainstorm reasons not to smoke. Use the reasons to design an advertising campaign against smoking.

LESSON 4

Other Drugs to Avoid

You will learn . . .

- how drug abuse harms health.
- the effects of marijuana, depressants, and stimulants.
- the effects of narcotics, inhalants, and steroids.

Vocabulary

- **tolerance**, *D25*
- **overdose**, *D25*
- **withdrawal**, *D25*

Do you know how drug abuse can affect your health? Drug abuse harms your mind, body, and relationships. Knowing the facts about drug abuse can help you make responsible decisions.

How Drug Abuse Harms Health

You read in Lesson 1 that drug abuse is the use of an illegal drug or the harmful use of a legal drug. Drug abuse harms health in many different ways.

Addiction A person who abuses a drug can become dependent on the drug. When the person is dependent he or she is addicted. Addictions are very difficult to stop.

Tolerance Over time a person can build up a tolerance to a drug. **Tolerance** is a condition in which more of a drug is needed to get the same effect. This increases the risk of an overdose. An **overdose** is too large an amount of a medicine or drug. An overdose can cause serious illness or even death.

Mental and Physical Problems Drug abuse harms the mind and body. It makes it difficult to concentrate and do well in school and sports. Over time drug abuse can harm the mind and body permanently.

Social Problems A person who abuses drugs may lose the trust and respect of family and friends. Drugs can make it hard to meet new people and make new friends, too.

Legal Problems Many drugs are illegal. Buying, selling, or using these drugs can send a person to jail.

Cost Drugs are expensive. People who are addicted to drugs may steal money to buy drugs. They may commit other crimes for money, too.

Withdrawal People who have addictions and then stop using drugs may go through withdrawal. **Withdrawal** is the unpleasant reaction experienced when a drug is no longer taken. People going through withdrawal may shake or feel nauseous. They may see and hear things that aren't there. The symptoms of withdrawal can be very serious and can even cause death.

Write About It!

Write an Expository Report
Choose a drug. Write a news report explaining why a person might start using the drug. Then describe how the person could become addicted. Finally, describe what might happen when the person stops using the drug. Present your report to the class.

▲ **Healthful activities benefit your mind.**

Marijuana, Depressants, and Stimulants

▲ **Marijuana comes from a plant called hemp or cannabis.**

Marijuana, stimulants, and depressants are drugs that people commonly abuse. *Marijuana* (mehr•uh•WAH•nuh) is a drug that is made from the cannabis plant. Depressants are drugs that slow down body functions. *Stimulants* are drugs that speed up body functions.

Marijuana

Most people who abuse marijuana smoke it. Some eat it in food. Marijuana is illegal in most places.

People use marijuana because it makes them feel high. This is caused by a chemical in marijuana called THC. *THC* changes the way people think and feel. People can become dependent on marijuana.

People who abuse marijuana can have trouble learning and remembering things. People who smoke marijuana may be more likely to get lung cancer. Marijuana also speeds up the heart. It increases appetite and slows down reaction time.

Depressants

Some depressants are legal if a doctor prescribes them. A doctor might prescribe *tranquilizers* (TRAN•kwuh•ligh•zerz) to help relieve anxiety. *Barbiturates* (bar•BI•chuh•ruhts) are like tranquilizers, but they are even stronger. They make people sleepy.

These depressants slow reaction time. They reduce blood pressure and slow the heart. They also slow down the thinking part of the brain. It is very dangerous to take them with alcohol and can even cause death.

Some people abuse depressants to help them relax. They become dependent and can't relax without taking depressants.

Stimulants

Stimulants make a person feel more alert. When some stimulants wear off, a person can become very depressed. This can even lead to suicide. A person may keep taking the stimulant to avoid this feeling. He or she becomes dependent.

Many people consume caffeine (ka•FEEN) every day. Caffeine is a legal stimulant found in coffee, tea, some sodas, and chocolate. Caffeine helps people feel alert. It can speed up the heart. It might irritate a person's stomach lining. You can become addicted to caffeine.

Amphetamines (am•FE•tuh•meenz) are drugs that increase alertness. Some amphetamines are legal with a prescription. Some diet pills contain amphetamines. Amphetamines also can help people with some kinds of sleep disorders or attention problems.

Both legal and illegal amphetamines can be addictive. These drugs speed up the heart and raise blood pressure. Abusing amphetamines can cause an irregular heart beat and brain damage.

Cocaine (koh•KAYN) is an illegal stimulant made from leaves of the coca plant. Cocaine makes a person feel high, and then very depressed. People usually inhale it. *Crack* is a very strong form of cocaine that people smoke. A person can become addicted to either one after just one or two uses. Both can cause brain damage.

Ecstasy, or MDMA, is an illegal drug that is both a stimulant and a depressant. It can harm the heart and brain.

What is a stimulant?

CAREERS

Drug Counselor

Drug counselors are trained to help people who abuse drugs. They also help the families of people who abuse drugs.

Some counselors work with people one-on-one. Others work with groups of people at the same time. Such counselors work at private offices, in hospitals and clinics, or at companies that have drug abuse programs.

▲ Vapors from spray paint are harmful if inhaled.

Narcotics, Inhalants, and Steroids

Some drugs are neither stimulants nor depressants. These drugs include narcotics, inhalants, and steroids.

Narcotics

Narcotics (nahr•KAH•tiks) are drugs that slow down the nervous system. Some narcotics are very useful for treating pain. *Morphine* (MOR•feen) and *codeine* (KOH•deen) are two such narcotics. These drugs are both made from the opium poppy plant. These drugs must be prescribed by a doctor. They can slow down the heart and breathing and can be addictive. They are very dangerous if a person takes them with alcohol.

Heroin (HEHR•uh•wuhn) is an illegal narcotic made from morphine. It can cause mental problems, infections, and breathing problems. Heroin is highly addictive. People who try to quit using heroin may have pain, upset stomachs, and movements they can't control. They may need medicine to help them handle the withdrawal symptoms.

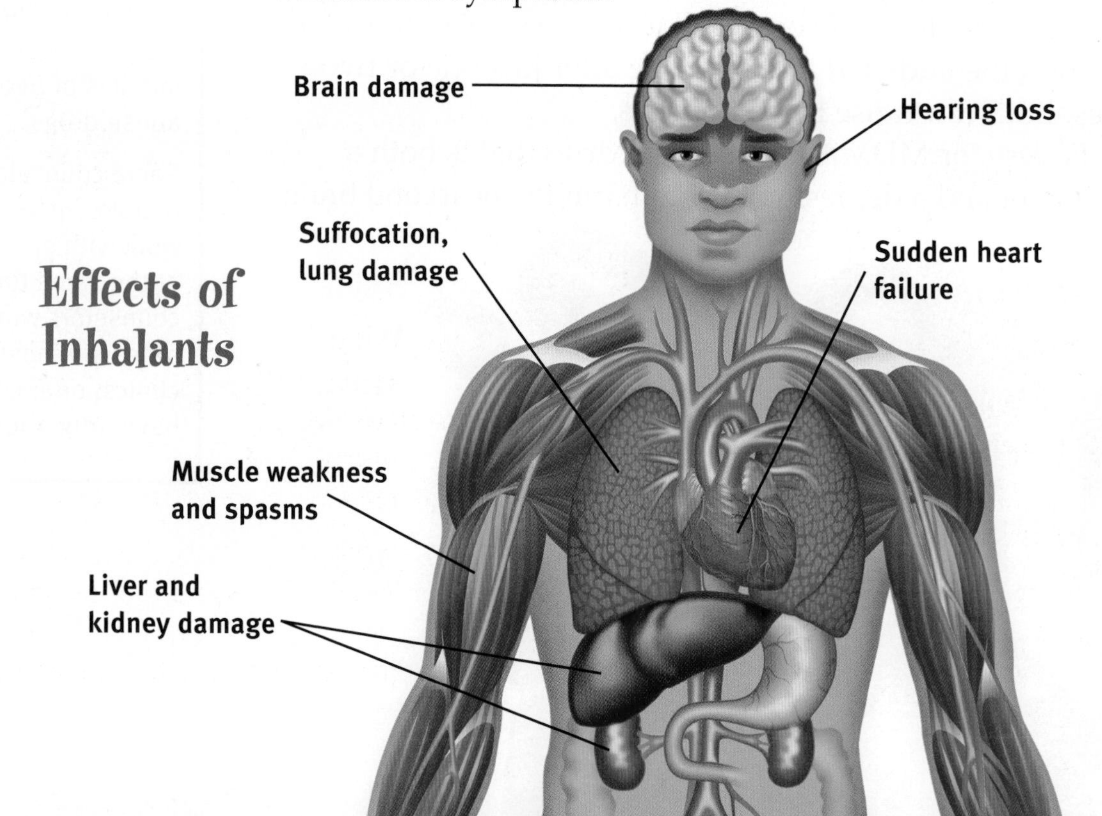

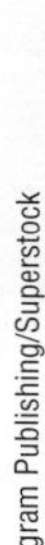

Inhalants

A chemical that is breathed is an *inhalant*. Gasoline, cleaners, paint, and some kinds of glue have harmful fumes. Some people breathe these fumes on purpose. The fumes make them feel high for a short time. Inhaling these chemicals can cause "sudden sniffing death." The heart suddenly stops beating, causing death.

Steroids

Steroids (STEHR•oydz) are drugs that act like hormones. They are legal with a prescription. Legal steroids help reduce swelling. They also can help people with asthma and other breathing problems.

Anabolic (AN•uh•bah•lik) steroids are similar to a male hormone. They can increase the size of muscles. Some people abuse these steroids. Steroids can damage the liver, heart, and kidneys. They can cause acne. They can harm the reproductive system.

Why do people abuse steroids?

ACTIVITY

LIFE SKILLS

CRITICAL THINKING

Manage Stress

Some people abuse drugs to escape from strong feelings caused by stress. But when the drugs wear off, the people still have the same feelings. Work with a partner to list healthful ways to manage stress without using drugs.

1. **Identify the signs of stress.** List ways you know you are stressed.
2. **Identify the cause of stress.** List some situations that can cause stress.
3. **Do something about the cause of stress.** Brainstorm healthful ways to manage each situation on your list.
4. **Take action to reduce the harmful effects of stress.** Write a list of ways to stay healthy when these situations occur.

Together with your partner, make a poster or sign listing your suggestions. Post it in your classroom or school.

LESSON REVIEW

Review Concepts

1. **List** four effects of drug abuse.
2. **Describe** the effects of marijuana, depressants, and stimulants.
3. **Describe** the effects of narcotics, inhalants, and steroids.

Critical Thinking

4. **Categorize** List at least two drugs that can harm each of these body systems: circulatory system, nervous system, and respiratory system. Describe what effect each drug has on the system.
5. LIFE SKILLS **Manage Stress** A friend is thinking about using marijuana to reduce stress. List five ways she could manage stress without using drugs.

LESSON 5

When Someone Abuses Drugs

You will learn . . .

- some reasons people abuse drugs.
- what help is available to people who abuse drugs and family members of people who abuse drugs.

Vocabulary

- recovery program, *D34*

People start abusing drugs for many different reasons. But once they start, it can be difficult to stop. A person who is addicted may need help to stop. Family and friends can support the person. The family and friends may need help, too.

Reasons People Abuse Drugs

People of all ages may abuse drugs for many different reasons. The following are reasons why some young people abuse drugs. Understanding these reasons can help you avoid drug abuse.

Family Difficulties Being part of a healthful family helps young people feel loved and supported. Young people who don't have this connection may feel empty. They want this empty feeling to go away. They might abuse drugs to change the way they feel. Help protect yourself by building strong relationships with your family.

Drug Abuse in the Family When adult family members abuse drugs, young people might do the same thing. They see adults they trust abuse drugs. They might believe it is okay to abuse drugs in the same situations. A person whose family member abuses drugs can find help by talking to other adult family members or to a responsible adult outside the family.

Peer Pressure Young people might believe that they need to abuse drugs to fit in with their peers. The pressure to feel accepted can feel overwhelming. Use the resistance skills you have learned to resist this negative peer pressure. Choose friends who are drug free.

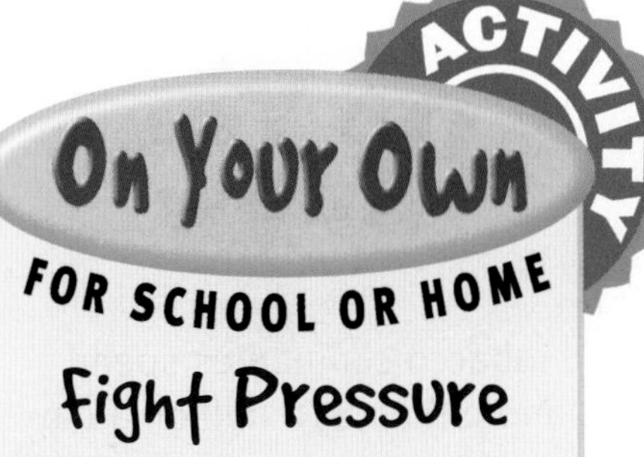

Fight Pressure

Talk to your parents or guardian about ways to handle peer pressure. Ask your parent or guardian for ways he or she has resisted pressure in the past. Role-play what to do if a peer pressures you to abuse drugs.

▲ **Your friends can influence you to do positive things together. How can you influence your friends not to abuse drugs?**

Tobacco companies spend more than $10 million a day on ads.

Negative Self-Concept Some young people do not think highly of themselves. They might feel awkward in social situations or have a negative body image. They believe they will feel better if they lose weight or have larger muscles. They are at a higher risk for abusing amphetamines, steroids, or other drugs. Developing a healthy self-concept can help protect you from this risk.

Depression Some young people have depression that does not go away. They want to escape their sad feelings. They might abuse drugs to change their mood. If you feel depressed, talk to a parent or guardian. The adult can help you get treatment if you need it.

Media Influence Companies that make tobacco and alcohol products spend millions of dollars on ads. Some young people are influenced by these ads. They believe the appeals in the ads instead of focusing on the harmful effects of the drug being advertised. Use your health knowledge when you see these ads. Don't let the appeals fool you into forgetting the health risks of tobacco and alcohol.

In addition to ads, tobacco and alcohol products are often shown in movies and television programs. They are often shown being used by film stars who make their use seem glamorous or fun. However, there is nothing glamorous or fun about the harmful effects of tobacco and alcohol.

Give one reason people start using drugs.

◀ **A parent or guardian can help you solve your problems without drugs.**

Getting Help

People who abuse drugs need help. Suppose that you have a friend who abuses drugs. You may think you are tattling on your friend if you tell a responsible adult. You're not. You're following laws and helping your friend. You are showing that you care about your friend's health.

If you know someone who abuses drugs, don't try to talk to the person about it. The person may become angry. He or she may feel like you are accusing him or her of doing something wrong. Drug abuse can affect how the person thinks and acts. He or she may react in ways that are dangerous to you. It is safer to talk to a responsible adult. The adult may be able to talk to the person more easily than you can.

When talking with a responsible adult, use I-messages to express your feelings. The adult may know how to find help for the person who is using drugs. He or she can also help you manage your feelings about the drug abuse.

Tell three people who could help a person who is abusing drugs.

Getting Help

Where to Get Help	What to Do
Home	• Talk to your parents or guardian. Ask for help.
School	• Your teacher, school nurse, or school counselor can help you. • Your school may have a special drug counselor as well.
Community	• Responsible adults such as members of the clergy can support you. • Many hospitals and clinics have programs to help people stop abusing drugs. • Special counselors help people who abuse drugs. Counselors also help family members of people who abuse drugs.

Write About It!

Write a Persuasive Slogan
Write a slogan to encourage people who abuse drugs to get help. Think of a short sentence or phrase. The slogan should be catchy and positive. With a group, perform your slogan as a cheer for your classmates.

List Reasons to Respect Yourself

Respect Identify reasons you respect yourself. Turn each one into a reason you could give for not using drugs. For example, if you respect yourself because you take care of your younger sister, you could say, "I don't want to use drugs because I wouldn't be able to think clearly enough to care for my sister." Practice using your reasons with a partner.

Recovery

A person who abuses drugs or alcohol may resist help if other people suggest it. The person who is addicted has to make the decision to get help on his or her own.

When a person decides to quit, he or she can enter a recovery program. A **recovery program** is a group that supports people who are trying to change. Programs such as *Alcoholics Anonymous* and *Narcotics Anonymous* help people stop abusing alcohol and other drugs. Sometimes people use prescription medicines to help with the physical effects of withdrawal.

Family Support

Living with a person who abuses drugs or alcohol can be hard. Sometimes the person may abuse family members. He or she may say and do hurtful things. Usually people who abuse drugs or alcohol don't know the harm they are causing their families.

Family members can show that they care by being supportive when the person decides to quit.

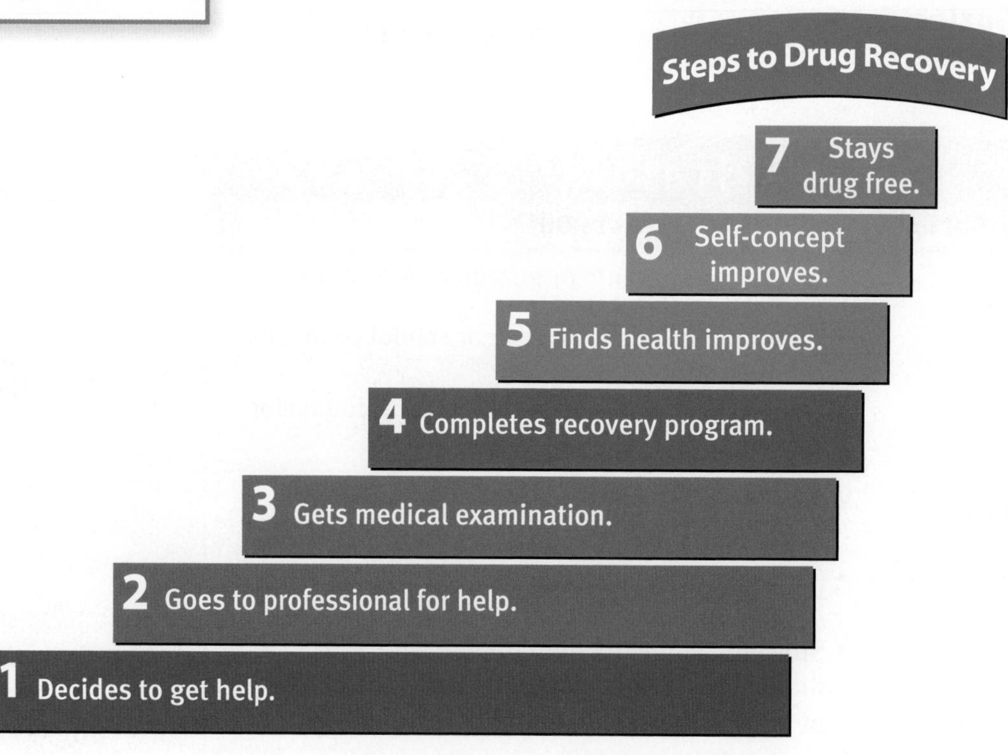

▲ **These are the steps a person follows when withdrawing from drugs.**

Sometimes family members need help. Counselors can help family members deal with the anger and pain they may feel because of the person's behavior. They may also help family members support a person who has decided to quit. Programs called *Nar-Anon* and *Al-Anon* help people who are close to people who have addictions. *Alateen* is a recovery program just for teens who are close to people who have addictions.

What do recovery programs do?

Counselors can help family members of people who abuse drugs.

ACTIVITY — LIFE SKILLS — CRITICAL THINKING

Use Communication Skills

Suppose you have a friend who abuses drugs. Work with a partner to role-play how to get help. Your partner should play a responsible adult.

1. **Choose the best way to communicate.** Would you talk to the responsible adult in person? On the phone? How else might you communicate?
2. **Send a clear message. Be polite.** Use I-messages to discuss the problem.
3. **Listen to each other.** The person playing the responsible adult should listen to the problem and offer suggestions.
4. **Make sure you understand each other.** Repeat what you hear. Make sure each of you understands what the other has said.

After you finish, switch roles and try again.

LESSON REVIEW

Review Concepts

1. **List** reasons why some people abuse drugs.
2. **Identify** three resources that can help people recover from drug abuse.

Critical Thinking

3. **Analyze** Why would a person with a negative self-concept be more likely to abuse drugs?
4. LIFE SKILLS **Use Communication Skills** How could you use communication skills to talk to a teacher about a friend who might be using drugs?

LESSON 6

Resisting Pressure

What you will learn...

- how to find ways to resist pressure to use drugs.
- some important laws and rules about tobacco, alcohol, and drugs.

Vocabulary

- drug free, *D37*

***Respect* is "to treat others with dignity and consideration." It's important to respect yourself, as well as other people. You show that you respect yourself when you say "no" to drug abuse.**

Saying "No!"

Peers or others may pressure you to use drugs. You can stay drug free even when you feel peer pressure. Being **drug free** is choosing not to abuse drugs.

Remember the resistance skills you have learned. Say "no" when someone pressures you to abuse drugs. Say it loudly and firmly. Give your reasons. If you need a good reason, choose one of the reasons below. Then walk away from the person. Be proud of yourself when you say "no" to drug abuse. You're respecting yourself and making a responsible decision.

What is being drug free?

▼ **Here are some good reasons not to abuse drugs. What other reasons can you think of?**

Aldo Murillo/Getty Images

Laws and Regulations on Drug Use

Choose the FDA, the DEA, or the ATF. Research the agency you choose. Find out what kinds of regulations the agency enforces. Use the e-Journal writing tool to summarize your findings in a report.

It's against the law to use many drugs. Some drugs that are legal for adults are not legal for people your age. You can go to jail if you break drug laws.

Three U.S. Government departments make policies and enforce laws about drugs. These are the Food and Drug Administration (FDA), the Drug Enforcement Administration (DEA), and the Bureau of Alcohol, Tobacco, and Firearms (ATF).

The FDA has *regulations*, or rules, about legal drugs such as amphetamines and codeine. The regulations tell doctors how to prescribe the drugs safely. The DEA enforces laws about illegal drugs. This agency works to stop people from bringing drugs into the United States. The agency also tries to stop the sale of illegal drugs.

The ATF deals with laws about alcohol and tobacco. This agency also enforces laws about guns.

Police officers can give you information about resisting drugs.

Christopher Kerrigan/McGraw-Hill Education

Your state and community have laws about drugs, too. These laws are there to protect people. Say "no" to drugs and you will not risk going to jail because of these laws. You will be making the responsible decision by saying "no."

Why are there laws about drug use?

▶ **Say "no" to drug abuse. Show that you respect yourself.**

ACTIVITY LIFE SKILLS CRITICAL THINKING

Use Resistance Skills

With a small group, write a skit showing different situations in which a young person refuses drugs. Make sure your skit includes these steps.

1. **Look at the person. Say "no" in a firm voice.**
2. **Give reasons for saying "no."** Include several different reasons in your skit.
3. **Match your behavior to your words.** Show different ways a person could resist pressure to use drugs.
4. **Ask an adult for help if you need it.** One person in your skit should play a responsible adult who helps the young person avoid drugs.

Perform your skit for your class.

LESSON REVIEW

Review Concepts

1. **Explain** how you can use resistance skills to say "no" to drug abuse.
2. **Describe** the roles of the FDA, DEA, and ATF.

Critical Thinking

3. **Analyze** A friend tells you that another friend pressured her to smoke marijuana. What would you recommend that your friend do?
4. LIFE SKILLS **Access Health Facts, Products, and Services** How could you find out more about the FDA's regulations about painkillers?
5. LIFE SKILLS **Use Resistance Skills** List five reasons you can give for saying "no" to drugs.

CHAPTER 7 REVIEW

Use Vocabulary

depressant, *D11*
drug, *D5*
drug misuse, *D8*
nicotine, *D17*
overdose, *D25*
recovery program, *D34*
secondhand smoke, *D20*
tar, *D17*

Choose the correct term from the list to complete each sentence.

1. A drug that slows down body functions is a(n) __?__.
2. A substance that changes how the mind or body works is a __?__.
3. If a person takes too much of a medicine or drug, that is a(n) __?__.
4. Exhaled smoke or the smoke from the end of a cigarette, cigar, or pipe is called __?__.
5. A stimulant drug found in tobacco is __?__.
6. A gummy substance found in tobacco smoke is __?__.
7. The accidental unsafe use of a medicine is __?__.
8. A group that supports people who are trying to change is a __?__.

Review Concepts

Answer each question in complete sentences.

9. How does nicotine affect the heart?
10. Explain how alcohol affects the liver.
11. What is heroin? How does it harm health?
12. Explain what it means to be drug free.
13. What resources are available for family members of people who abuse drugs or alcohol?
14. Summarize the effects of steroids on the body.
15. Identify five reasons a person may abuse drugs.
16. Describe what role the FDA, DEA, and ATF play in preventing drug use.

Reading Comprehension

Answer each question in complete sentences.

Exhaled smoke and smoke from the burning end of a cigarette, cigar, or pipe is called secondhand smoke. Secondhand smoke is a *carcinogen* (kar•SIN•uh•juhn). A carcinogen is something that can cause cancer. About 3,000 nonsmokers die every year from lung cancer caused by secondhand smoke. Secondhand smoke can cause heart disease. Young people who breathe secondhand smoke have a higher risk for lung problems. They get asthma more often. They also get more lung infections, such as pneumonia and bronchitis.

17. What is secondhand smoke?
18. What problems can secondhand smoke cause?
19. What is a carcinogen?

Critical Thinking/Problem Solving

Answer each question in complete sentences.

Analyze Concepts

20. A person who is addicted to drugs decides to get help. What are the steps in drug recovery that the person will go through?

21. A friend offers you an alcoholic drink. What five reasons could you give for saying "no" to your friend?

22. How do advertisements try to influence the choices you make about alcohol use? How can you protect yourself from those influences?

23. Why is it important to know about the toxins in tobacco smoke?

24. Your doctor gave you a new medicine. Now you have a headache and a rash. What do you think is happening? What should you do?

Practice Life Skills

25. **Be a Health Advocate** Your friend started smoking. She says she feels cool now. What would you tell your friend to encourage her to quit?

26. **Make Responsible Decisions** You are in class and feel very sleepy. A classmate offers you a medicine that helps him feel awake. What should you do? Use the *Guidelines for Making Responsible Decisions*™ to help you decide.

Read Graphics

Use the graph to answer each question in complete sentences.

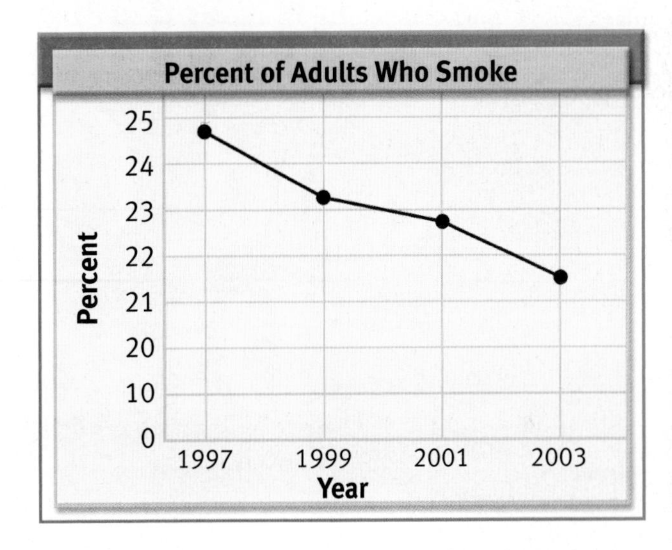

27. Since 1997, has the percent of adults who smoke increased or decreased?

28. Approximately how much did the number of smokers change from 1997 to 2001?

29. Based on the graph, do you think the percent of adults who smoke will increase or decrease in the next five years? Why?

CHAPTER 8

Communicable and Chronic Diseases

Lesson 1 • Communicable Diseases *D44*

Lesson 2 • How Your Body Fights Infection ... *D50*

Lesson 3 • Signs of Illness *D56*

Lesson 4 • Chronic Disease and the Heart *D60*

Lesson 5 • Chronic Disease: Cancer *D66*

Lesson 6 • Other Chronic Diseases........ *D72*

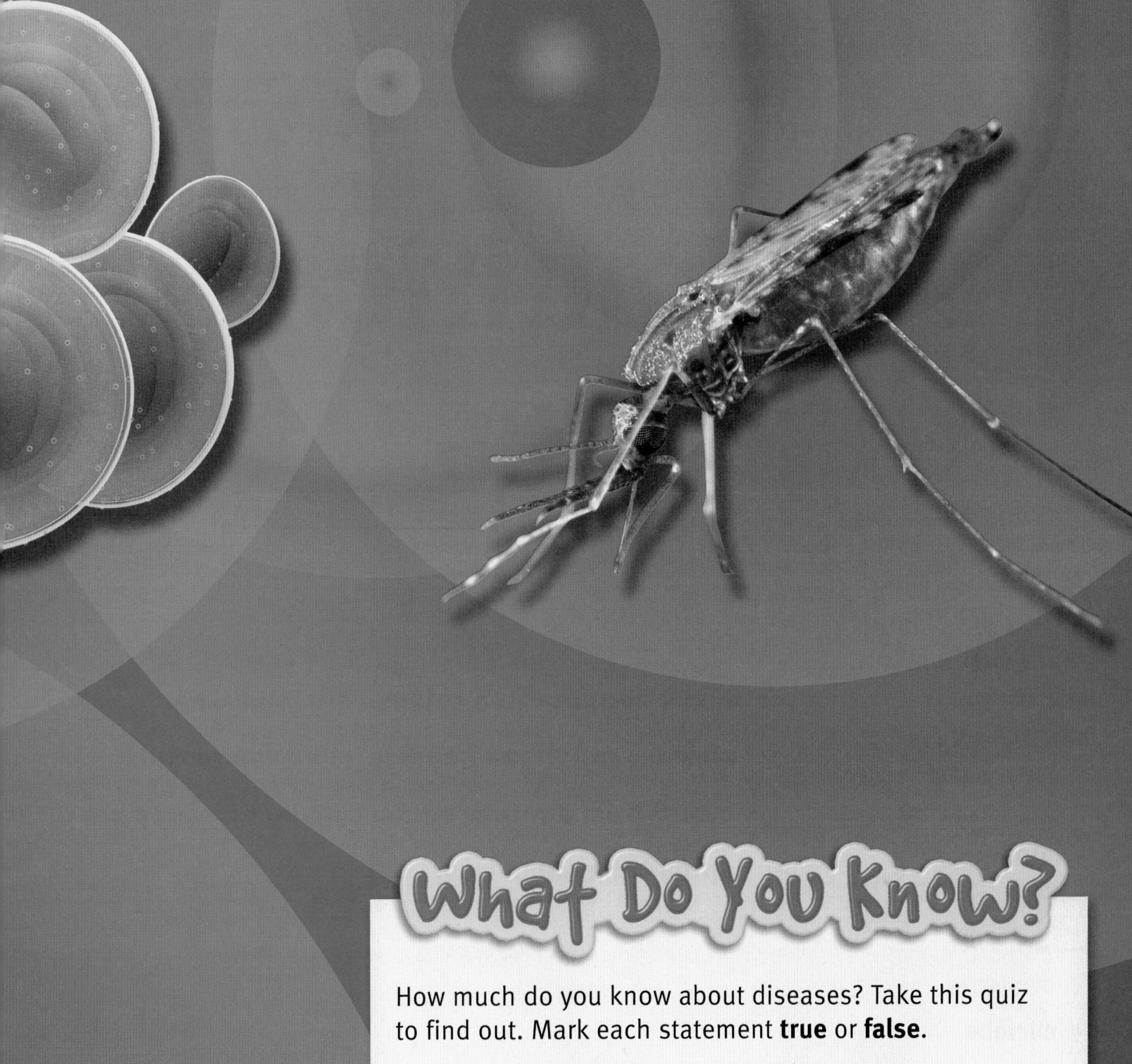

What Do You Know?

How much do you know about diseases? Take this quiz to find out. Mark each statement **true** or **false**.

1. You can catch cancer from another person.
2. What you eat affects your risk of getting some diseases.
3. Washing your hands can help prevent illness.
4. Vaccines contain pathogens.
5. Diabetes affects children only.

If you marked statements 2, 3, 4 **true** and statements 1 and 5 **false**, you already know a lot about disease! To find out more read **Communicable and Chronic Diseases**.

Communicable Diseases

You will learn . . .

- what communicable and noncommunicable diseases are.
- how pathogens spread and cause disease.
- how pathogens get into your body.

Vocabulary

- **microbe**, *D46*
- **virus**, *D46*
- **bacteria**, *D46*
- **fungus**, *D46*
- **protozoa**, *D46*

Many people catch colds every year. They sneeze, cough, and sniffle. It's easy to catch a cold from another person. Many diseases are spread the same way.

Types of Disease

People in the United States catch more than 60 million colds every year. There are another 95 million cases of flu. Colds and flu are communicable diseases. A *disease* is a condition that weakens or harms part of the body. *Communicable diseases* are diseases that spread from person to person. Colds, flu, and chicken pox are examples of communicable diseases. Many communicable diseases can be treated and cured.

Pathogens, sometimes called germs, cause communicable diseases. People, objects, and animals can spread pathogens. Pathogens are too small to see with your eyes. This means you can spread them without knowing it.

Noncommunicable diseases are diseases that you can't catch from another person. Heart disease, cancer, and diabetes are noncommunicable diseases. Some noncommunicable diseases can be treated and even cured. Others can last a lifetime.

Some pathogens are so small, you could fit 50 million of them into one meter!

Give one example each of a communicable disease and a noncommunicable disease.

▶ A cold is one kind of communicable disease. What others can you think of?

Tom Le Goff/Getty Images

Pathogens

A **microbe** (MIGH•krohb), or *microorganism* (migh•kroh•AWR•guh•niz•uhm), is a living thing that is too small to be seen without a microscope. Pathogens are microbes. The four kinds of pathogens that cause communicable diseases are viruses, bacteria, fungi, and protozoa.

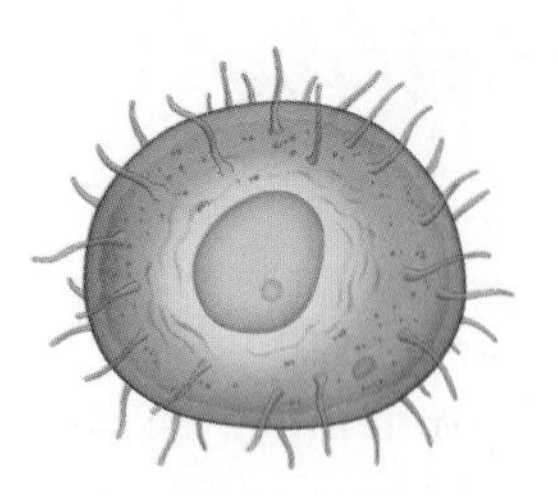

▲ Cocci

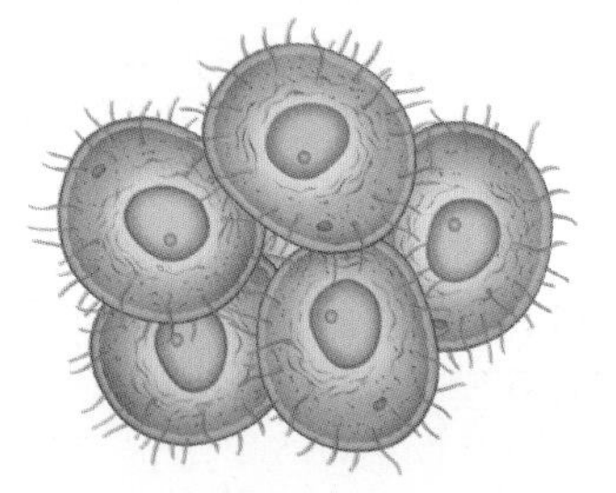

▲ Cluster of cocci

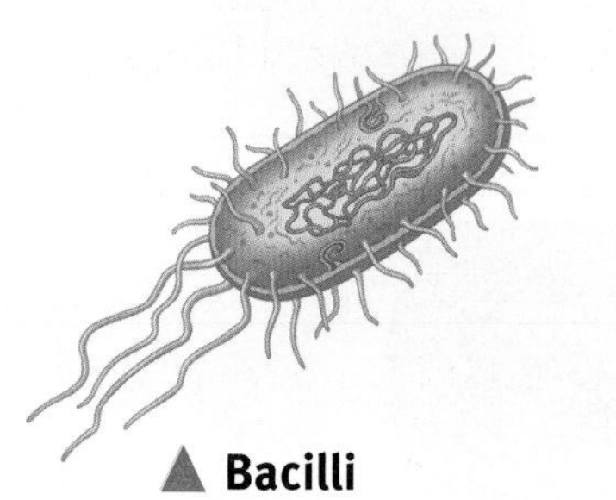

▲ Bacilli

▲ Chain of bacilli

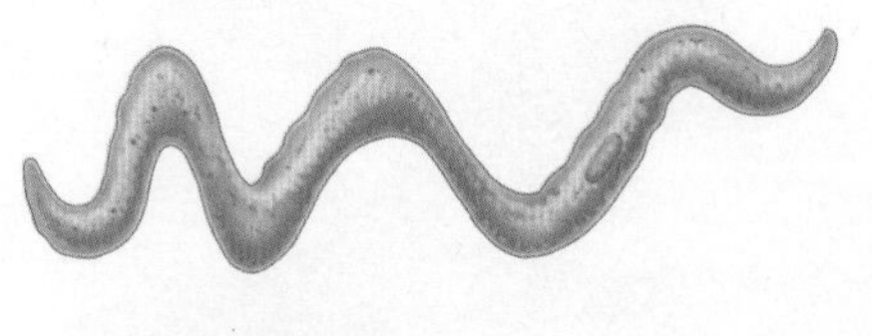

▲ Spirilla

- A **virus** (VIGH•ruhs) is a very tiny particle that makes copies of itself inside a cell. These copies spread to other cells. The viruses make toxins that cause disease. Viruses cause colds, flu, and measles, for example.
- **Bacteria** (bak•TEER•ee•uh) are the smallest living microbes that can reproduce on their own. They don't have to be inside a cell. Some bacteria live in your body. These bacteria help break down food. Other bacteria can cause disease if they get into your body. Bacteria cause such diseases as strep throat, some kinds of pneumonia, and many ear infections.
- A **fungus** (FUHNG•guhs) is a diseased, spongy growth on the body. It is actually a plantlike living thing. The plural of fungus is *fungi* (FUHN•jigh). Fungi cause ringworm and athlete's foot. These diseases cause skin rashes.
- **Protozoa** (proh•tuh•ZOH•uh) are simple one-celled microbes that are much larger than bacteria. Malaria is one disease caused by protozoa. Malaria is rare in the United States. Mosquitoes in other parts of the world carry the protozoa that cause malaria. Malaria causes fever and flu-like symptoms.

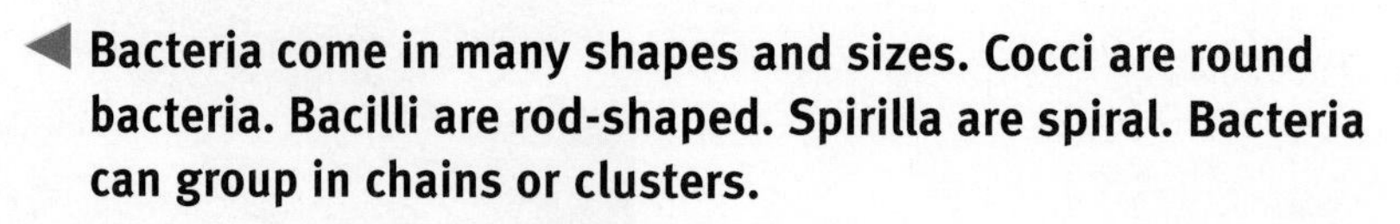

◀ **Bacteria come in many shapes and sizes. Cocci are round bacteria. Bacilli are rod-shaped. Spirilla are spiral. Bacteria can group in chains or clusters.**

Pathogens on the Move

Pathogens are all around you. They spread easily by contaminating objects, the air, and food. Here are examples of how this contamination can occur.

Suppose someone sneezes near a doorknob. Tiny droplets fly out of his or her nose and land on the doorknob. When you touch the doorknob, viruses move from the doorknob to your hand. You then touch your nose or bite your fingernails. The viruses could get into your body.

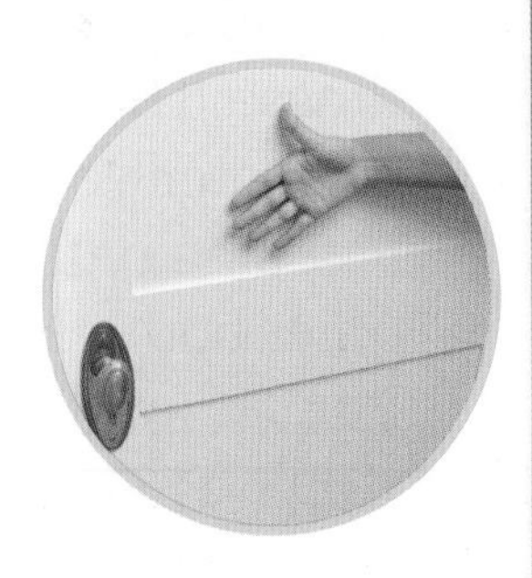

A person who has a cold coughs. Viruses fly out into the air. You're standing next to the person. You could breathe in the viruses.

Suppose an insect gets pathogens on its body as it crawls on garbage. Later it lands on a sandwich you're eating. The insect spreads pathogens to your food. People with unclean hands can also spread pathogens to food. A person who eats the food can become ill.

Pathogens can enter a water system through leaks or breaks in pipes. Dirt or waste with pathogens can get into the water. This water must be boiled or treated with chemicals to kill the pathogens. It's safer not to drink from any stream or pipe outside. It's also best not to swim in polluted lakes.

Ticks, mosquitoes, and fleas can carry pathogens in their bodies. When they bite a person or an animal, they can spread the pathogens. Dogs, cats, and wild animals can also spread pathogens through bites or scratches.

List the four main types of pathogens.

BUILD Character ACTIVITY

Plan to Prevent Disease

Citizenship Many pathogens grow in garbage. Insects can carry the pathogens from the garbage to people. You can help your community by throwing away your trash. Work with a partner. List three things you could do to dispose of garbage in a healthful manner, such as throwing litter into a trash can. Next to each idea, describe how it would help prevent disease.

How Pathogens Enter the Body

ACTIVITY

Art LINK

Make a Tissue Box

Cover a tissue box with paper. Decorate the box with pictures that show how to prevent the spread of disease. Use the box to hold tissues at home or in school.

Pathogens must enter your body before they can cause illness. The diagram below shows some ways pathogens can enter your body.

Some pathogens enter your body through blood or other body fluids. One example of these pathogens is human immunodeficiency virus, or HIV. HIV causes AIDS. This virus can spread when the body fluids of a person with HIV get into another person's body, such as through a cut or break in the skin.

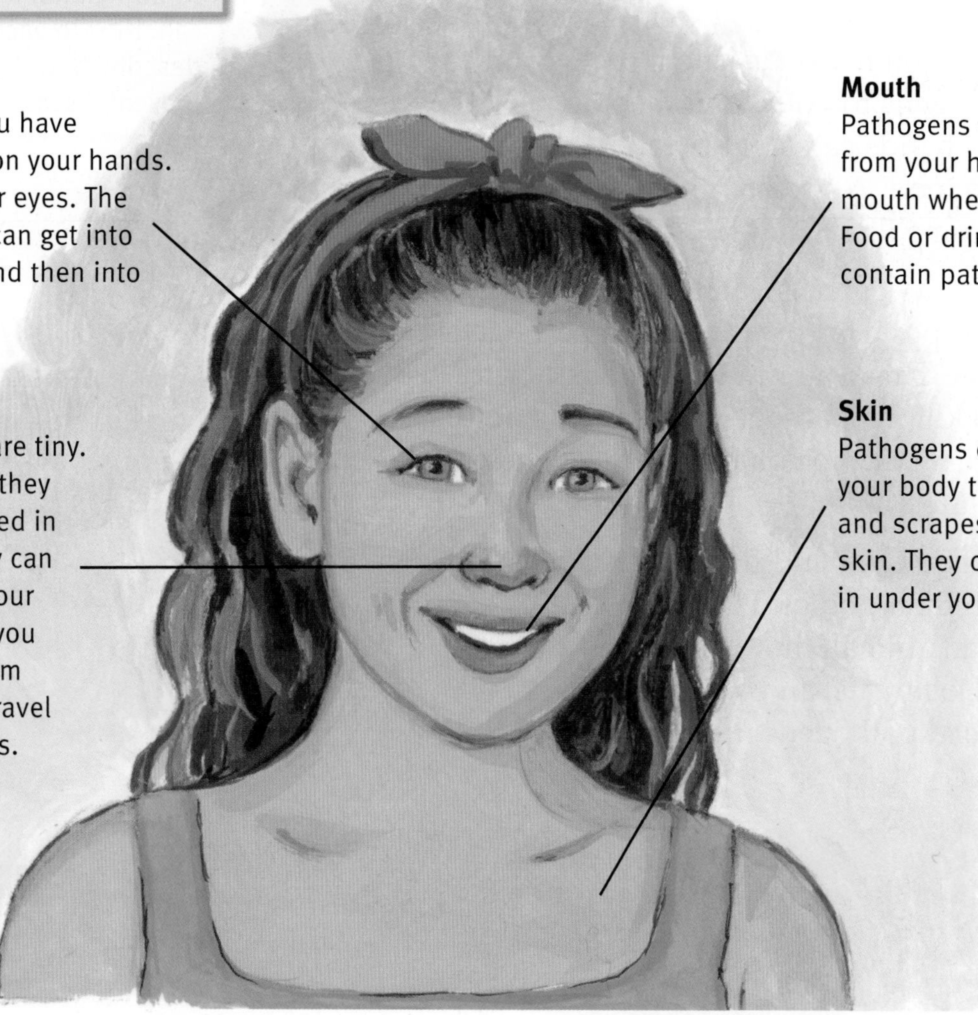

Eyes
Suppose you have pathogens on your hands. You rub your eyes. The pathogens can get into your eyes and then into your body.

Nose
Pathogens are tiny. This means they can be carried in the air. They can move into your nose when you breathe. From there they travel to your lungs.

Mouth
Pathogens can move from your hands to your mouth when you eat. Food or drinks can also contain pathogens.

Skin
Pathogens can enter your body through cuts and scrapes in your skin. They can also get in under your nails.

Reducing Risk

Here are some healthful tips for reducing your risk of getting communicable diseases.

- **Wash your hands often** with soap and water. Soap and water can remove many germs from your hands.
- **Don't share** cups, eating utensils, or bottles. You can pick up another person's pathogens.
- **Keep your fingers and other objects out** of your eyes, nose, and mouth. Pathogens that get into these body parts can cause disease.
- **Cover your nose and mouth** with a tissue when you cough or sneeze.
- **Get a flu vaccine** if a doctor recommends it. There is a new vaccine every year. The vaccine protects you against the flu viruses that are most common that year.

How can pathogens enter your body through your skin?

ACTIVITY — LIFE SKILLS — CRITICAL THINKING

Practice Healthful Behaviors

1. **Learn about a healthful behavior. Myth** Hand washing is just a way of making your hands look better. **Fact** Hand washing is an important way to stop disease from spreading. Gather information about good hand-washing techniques. Work with an adult to find tips on the Internet or ask the school nurse.
2. **Practice the behavior.** Wash your hands after you use the restroom, play with a pet, or touch something handled by a person who is ill. Wash your hands before you prepare or eat food.
3. **Ask for help if you need it.** Ask your parents or guardian for ways to remember to wash your hands. Make a list of the ideas they suggest.
4. **Make the behavior a habit.** Create a colorful sign. List and draw the steps for hand-washing. Use the list of tips to think of places where you could put the sign. Post it where it will help you and your family remember to wash your hands.

LESSON REVIEW

Review Concepts

1. **Distinguish** between communicable and noncommunicable diseases.
2. **List** five ways pathogens can spread.
3. **Explain** how pathogens can enter your body.

Critical Thinking

4. **Synthesize** Why is it important to wash your hands before you eat?
5. LIFE SKILLS **Practice Healthful Behaviors** List three behaviors that can help you prevent the spread of pathogens. Explain how you can make each behavior a habit.

LESSON 2

How Your Body Fights Infection

You will learn . . .

- what are the stages of disease.
- how the body fights illness.
- ways to help strengthen your body's immune system.

Vocabulary

- **cilia**, *D52*
- **white blood cells**, *D53*
- **fever**, *D53*
- **antibody**, *D53*
- **immunization**, *D53*

Vaccinations are shots that you may get at the doctor's office. Some people do not like to get these shots, but they are very important. They help your body fight disease. Vaccinations are just one way you can help your body build its defenses against pathogens.

Stages of Disease

Many pathogens enter your body. However, your body defenses quickly destroy the pathogens. You do not develop symptoms of disease.

Sometimes your body defenses do not destroy pathogens that invade your body. When this happens, you get the disease caused by the pathogens. The disease affects your body in stages.

Three Stages of Disease

Incubation The *incubation period* is the period from the time a pathogen gets into your body until you start showing signs of the disease. The incubation period may last from a few hours or days to a few weeks or longer. For example, the incubation period for chicken pox is 10–21 days. Flu takes 1–4 days. During this time your body tries to destroy the pathogens.

You can spread many communicable diseases during this time. The diseases are *contagious* (kuhn•TAY•juhs), or spreadable. You can spread chicken pox during the last day or two of the incubation period, for example.

Acute The *acute period* of a disease is the time during which you show symptoms. The length of the acute period varies. For chicken pox it usually lasts 5–10 days. The exact length depends on how well your body fights off the disease.

Most communicable diseases are contagious during the acute period. You can spread flu for 3–7 days or more after you start to show symptoms.

Recovery The *recovery period* is the time during which most of the symptoms go away, but you are not yet well. During this time you may still be able to spread the disease. If you take care of your body, you can have a short recovery period. If you try to do too much too fast, you can *relapse,* or go back to the acute period of the disease.

▼ **The earlier a disease is found and treated, the greater the chances of a quick recovery.**

What happens in the acute period?

The Immune System

CAREERS

Public Health Worker

The main job of a public health worker is to help protect and improve the lives of people and keep them safe from disease. Some public health workers keep track of communicable diseases that may be spreading in a community. Others work to keep public places and water supplies clean and healthful. Most public health workers work for local, state, or federal governments.

The *immune system* is the body system that helps protect your body from disease. It has two main lines of defense against pathogens. The first line of defense works against all kinds of pathogens. Other defenses work against particular pathogens.

First-Line Defenses

Many parts of your body work together to protect you from pathogens.

- **Your skin** forms a barrier between you and the environment. Unbroken skin blocks pathogens from entering your body.
- **Your air passages,** such as your nose and throat, are lined with tiny hairs called **cilia**. Cilia trap pathogens in the air you inhale so pathogens don't enter your lungs. Coughing or sneezing forces the pathogens out of your body.
- **Your eyes** produce tears. The tears coat your eyes when you blink. They wash away dust particles that may contain pathogens. Tears also contain chemicals that kill bacteria.
- **Your stomach** contains acids. These stomach acids can kill many pathogens that you swallow.
- **Your nose and throat** are lined with a moist coating called *mucus* (MYEW•kuhs). Some pathogens from the air are trapped in mucus. This keeps them from reaching your lungs. Many of these pathogens are released when you cough or blow your nose.

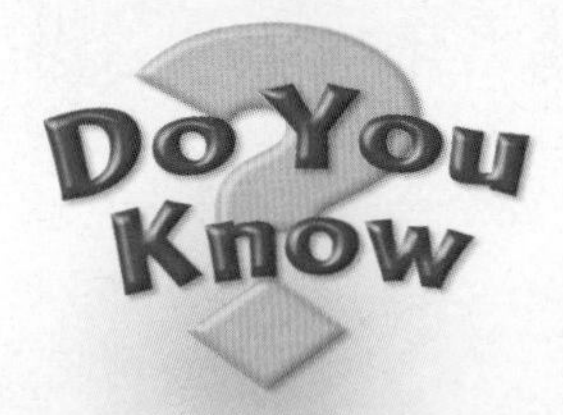

Myth Body piercing and tattoos are safe if done by a trusted friend.

Fact Any time you pierce the skin you risk infection with disease. A parent or guardian must approve the tattoo or body piercing. The procedure must be done by a licensed professional.

Other Body Defenses

If pathogens do get into your body, other defenses fight them. Three of these defenses are white blood cells, fever, and antibodies.

Your **white blood cells** are cells that surround and destroy pathogens. They attack pathogens that invade your body.

A **fever** is a rise in your body's temperature. It kills some pathogens and makes it harder for others to reproduce.

An **antibody** is a substance in blood that helps fight pathogens. Your blood contains many antibodies. Each antibody protects you from a different pathogen.

Suppose you have chicken pox. While you are ill, your body is making antibodies. The antibodies attack the virus that causes chicken pox. These antibodies stay in your blood for a long time. They may even stay for the rest of your life.

What happens if the virus that causes chicken pox gets into your body again? The antibodies destroy it. This keeps you from getting ill again. Now you are immune to chicken pox. **Immunization** is protection from getting a particular disease.

You can also become immune by getting a vaccine for a disease. Vaccines contain dead or weakened pathogens. There are vaccines for diseases including measles, mumps, polio, chicken pox, and flu. Many are given as shots. Some are liquids. You can get vaccines from a doctor or community health department.

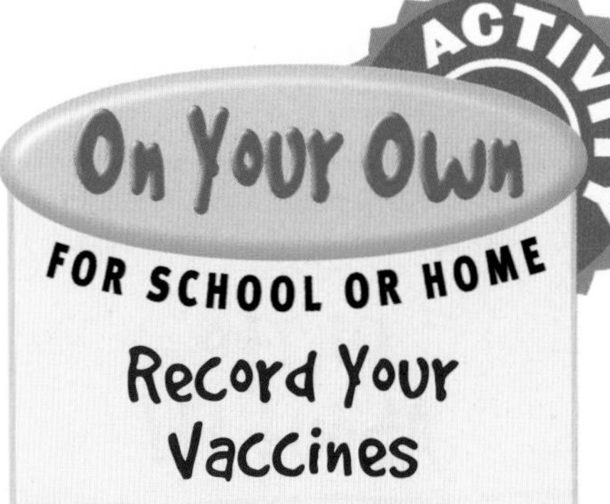

Record Your Vaccines

Work with a parent or guardian. Make an immunization record that lists all the vaccines you have received. Find out which vaccines are required in your school. In most places students are vaccinated before they start school.

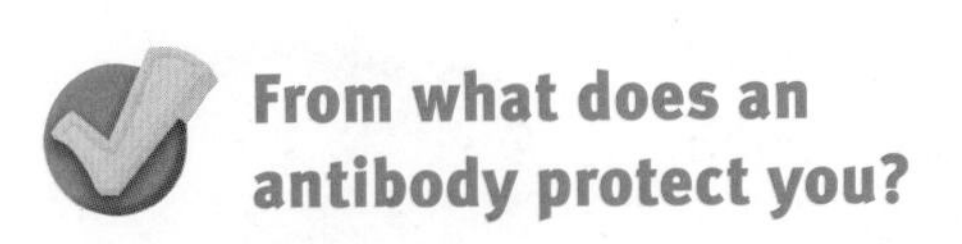

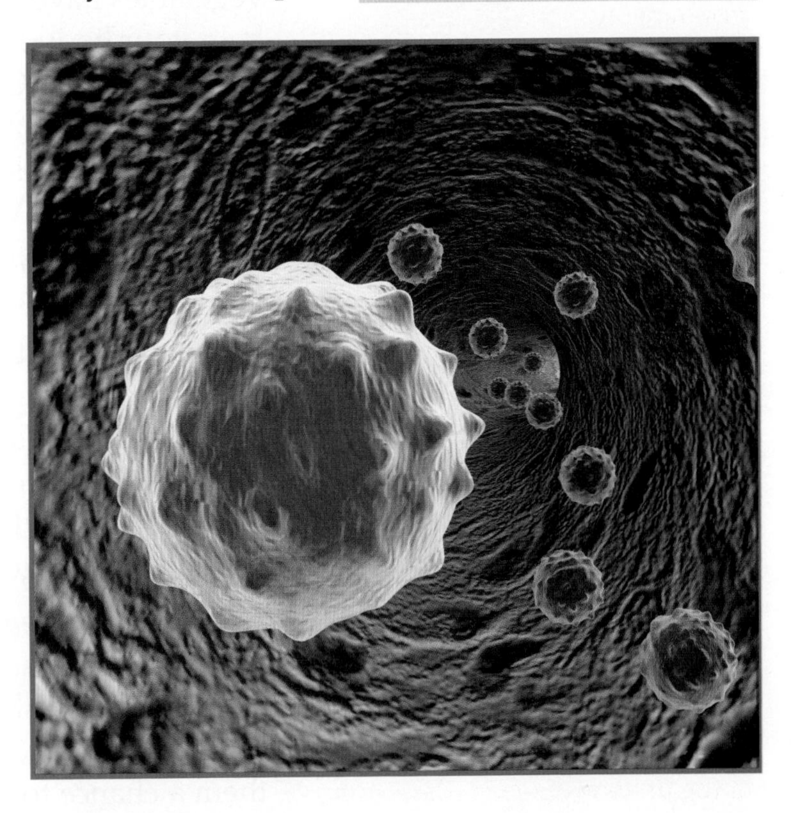

▲ White blood cells produce the antibodies that help your body fight disease.

Purestock/SuperStock

Keeping Your Immune System Strong

You can help your immune system do its job by keeping pathogens out of your body. Choose behaviors that help your immune system stay strong. Avoid behaviors that weaken your immune system.

Habits to Form

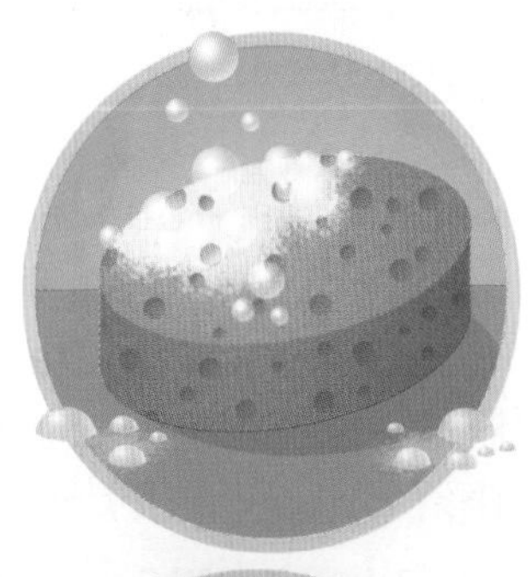

- **Keep yourself clean.** Bathe or shower every day. This removes pathogens from your skin. Wash your hands with soap and warm water. Wash and cover cuts and scrapes to keep pathogens from entering your body through the skin.

- **Eat healthful foods.** Fruits and vegetables have vitamins that help your body fight disease. Vitamins A and C help your immune system. Carrots, cantaloupe, and seafood are good sources of Vitamin A. Strawberries, oranges, and tomatoes all have Vitamin C.

- **Drink water.** Water helps keep the mucus in your nose and throat moist so that it can trap pathogens. Most people need about 6–8 cups of water daily, which includes water in beverages and foods.

- **Get enough sleep.** A person your age needs 8–10 hours of sleep each day. When you sleep, your body organs are less active. This gives them a chance to rest. Rest keeps them strong so that they can fight off infections.

Write About It!

Write a Reflective Essay
Think about last week. What did you do to help keep your immune system strong? What could you have done better? Write a short essay describing what you did and what you'd like to improve next week.

Habits to Avoid

Some people choose behaviors that can harm their immune systems. Avoid these behaviors to help protect your immune system.

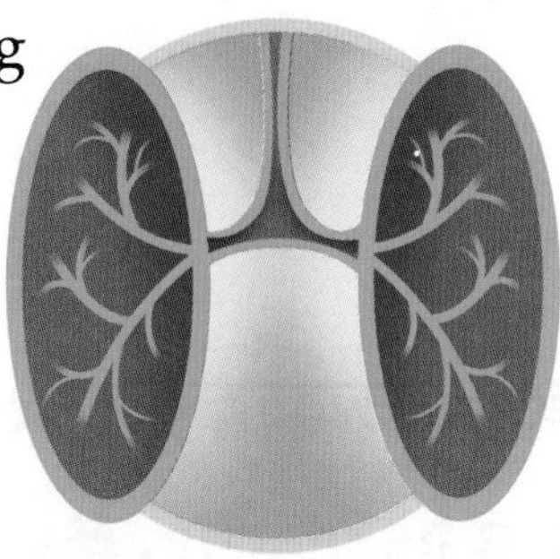

Do Not Smoke Smoking paralyzes the cilia that line the air passages. The cilia cannot move to trap pathogens that get into the lungs.

Don't Drink Alcohol Alcohol irritates the lining of the stomach. The stomach acids won't kill as many pathogens.

How does avoiding alcohol protect your immune system?

ACTIVITY LIFE SKILLS CRITICAL THINKING

Access Health Services

Many health clinics offer free or low-cost vaccines. Make a brochure showing where to find these services in your community.

1. **Identify when you might need health services.** Most people get vaccines as children. People need certain vaccines before they travel to other countries. Find out at what age vaccines are required. Make a list of which vaccines are given when.
2. **Identify where you might find health services.** List three resources you can use to find free or low-cost vaccines in your community.
3. **Find the health services you need.** With your parent or guardian find one clinic where vaccines are available. Find out when the clinic is open, who can use it, and which vaccines it offers. Find more than one clinic if you can. Make a chart showing the vaccines you listed in Step 1 and which clinics give those vaccines.
4. **Evaluate the health services.** Decide which clinics to include in your brochure. Then make a brochure listing the clinics, your chart, and other important facts about the clinic's services.

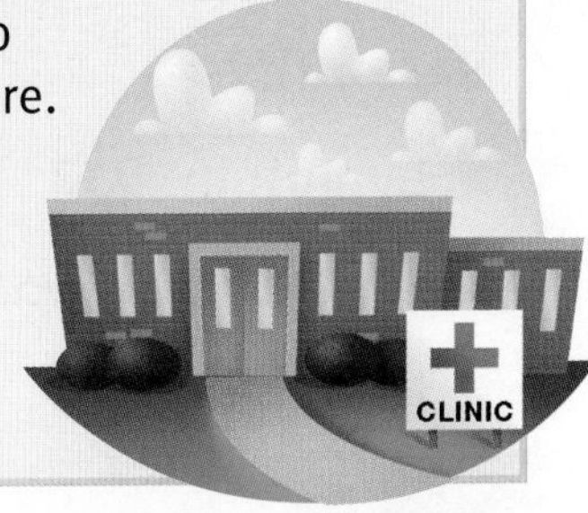

LESSON REVIEW

Review Concepts

1. **Describe** the three stages of disease.
2. **Explain** how the immune system works.
3. **List** six ways you can strengthen your immune system.

Critical Thinking

4. **Compare** How is receiving a vaccine similar to having a disease?
5. LIFE SKILLS **Access Health Facts** People who travel to foreign countries sometimes need vaccines against certain diseases. How could you find out which vaccines you need to travel to a country?

LESSON 3

Signs of Illness

You will learn . . .

- what symptoms are typical of common diseases.
- to identify common communicable diseases
- ways common communicable diseases can be treated

Vocabulary

- **symptom**, *D57*
- **strep throat**, *D58*
- **antibiotic**, *D58*

How do you know when you're ill? Maybe you wake up with a headache. Maybe you have a fever. Maybe you have a rash. Each of these symptoms may mean you have an illness.

Common Signs of Illness

A **symptom** is a change in your body's condition or function that may signify disease. Fever, body aches, and tiredness are common symptoms of disease. Symptoms that you can see, such as a rash, are also called *signs* of disease.

Many diseases have the same kinds of symptoms. Suppose that you have a cough. You might have a cold, the flu, or another disease. The doctor looks at all your symptoms. He or she may need to do medical tests. Then he or she can decide what illness you have.

Some symptoms are signs of a specific disease. A rash that looks like the bull's-eye on a target, for example, is a symptom of Lyme disease. The spots caused by chicken pox look different from the spots caused by other diseases.

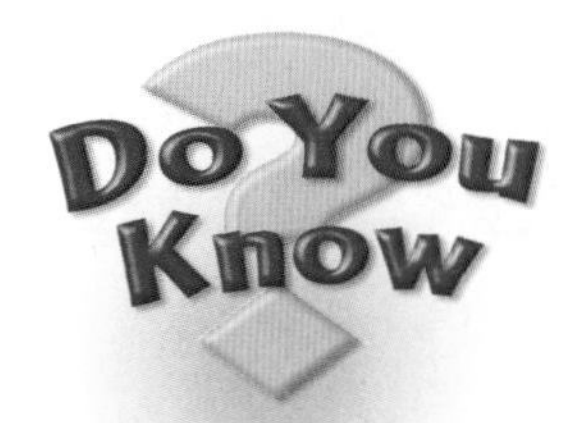

Normal body temperature is 98.6° Fahrenheit. Your normal temperature may be a little higher or lower. A temperature over 100°F means you have a fever.

What is a symptom?

Signs and Symptoms of Some Diseases

Illness	Signs and Symptoms
Cold	Tiredness, cough, sore throat, stuffy or runny nose, sneezing, watery eyes
Flu	Fever, tiredness, cough, headache, decreased appetite, body aches, chills
Strep throat	Sore throat, gray and white patches on throat, body aches, fever, loss of appetite, often swelling in the throat
Chicken pox	Fever, itchy skin rash that looks like small blisters or spots
Lyme disease	"Bull's-eye" rash, tiredness, headache, stiff neck, fever, chills, muscle and joint pain
West Nile virus	Fever, headaches, body aches, upset stomach. Serious cases can sometimes lead to shaking, numbness, and vision problems.

This rash is one of the first symptoms of Lyme disease. The disease is carried by ticks.

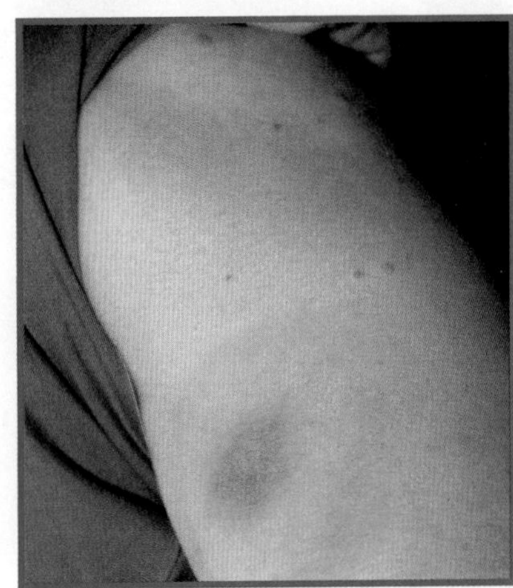

Common Communicable Diseases

Do You Know

In 2003 a new communicable disease appeared in Asia. Severe Acute Respiratory Syndrome, or SARS, causes high fever, coughing, and pneumonia. It's caused by a virus that scientists hadn't known about before. SARS has now largely been controlled.

▲ **Cover your nose when you sneeze to avoid spreading pathogens.**

The most common communicable diseases are colds, flu, and strep throat.

Colds and Flu

A *cold* is an infection of the respiratory system caused by a virus. There is no cure for a cold. A doctor may tell you to get extra rest, drink lots of water, and take over-the-counter medicines to ease your symptoms.

Flu, or influenza (in•flew•EN•zuh), is also caused by a virus. Flu can be treated, but not cured. Often people treat the flu the same way they treat a cold.

Sore Throats

Sore throats can be caused by bacteria or viruses. **Strep throat** is a bacterial infection of the throat. See a doctor if you show signs of this disease. Untreated strep throat can cause permanent heart damage.

To treat strep throat, a doctor may prescribe antibiotics. An **antibiotic** is a drug used to kill or slow the growth of germs that cause disease. Antibiotics fight only bacteria, not viruses. If your sore throat is caused by a virus, a doctor may tell you to rest, drink liquids, and take over-the-counter medicines to help relieve your symptoms. He will not prescribe antibiotics.

When Someone Is Ill

What can you do when someone you know is ill? You might offer to pick up a classmate's work from school. You could be extra quiet when an ill family member needs to rest. Just asking people how they feel shows care and concern.

What do antibiotics do?

ACTIVITY

LIFE SKILLS

CRITICAL THINKING

Make Responsible Decisions

A friend invites you to go camping. You've heard that there are many ticks that carry Lyme disease in the area. What should you pack? Make a list.

1. **Identify your choices.** Check them out with your parent or trusted adult. You could take insect repellent, long-sleeved shirts, and long pants. You could just take a T-shirt and shorts. Write down everything you might take.
2. **Evaluate each choice. Use the *Guidelines for Making Responsible Decisions™*.** Think about each choice. Cross out items that are not healthful choices.
3. **Identify the responsible decision.** Check this out with your parent or trusted adult. Write a packing list for the camping trip based on your decision.
4. **Evaluate your decision.** Next to each item, write down why you chose to include it.

Guidelines for Making Responsible Decisions™

- **Is it healthful?**
- **Is it safe?**
- **Does it follow rules and laws?**
- **Does it show respect for myself and others?**
- **Does it follow family guidelines?**
- **Does it show good character?**

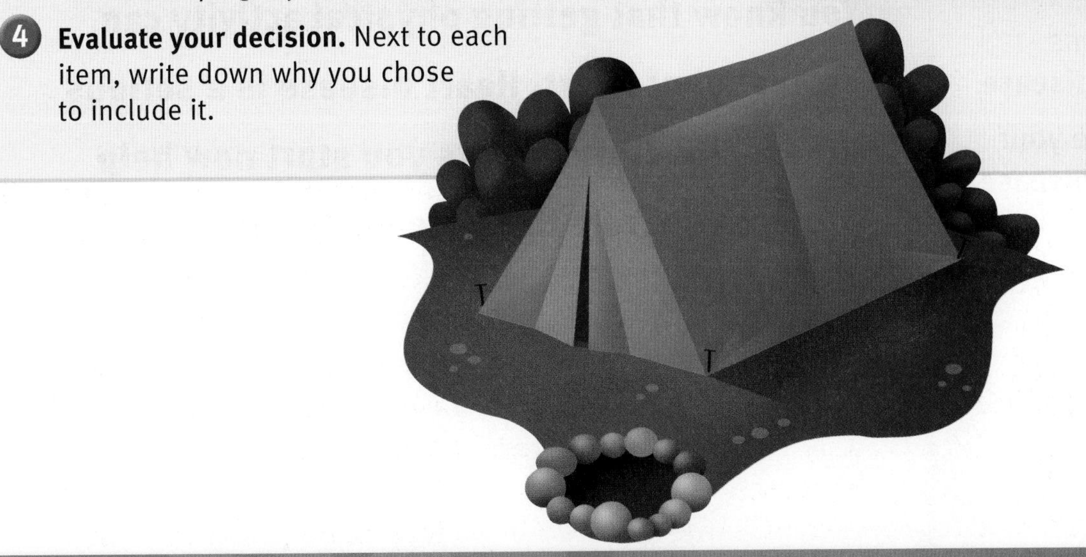

LESSON REVIEW

Review Concepts

1. **Explain** the difference between a sign and symptom of a disease.
2. **List** the symptoms of a cold, the flu, and strep throat.
3. **Explain** how colds, flu, and strep throat can be treated.

Critical Thinking

4. **Compare** How are strep throat and the flu alike?
5. LIFE SKILLS **Make Responsible Decisions** You have a bad cold. Your class is going on a field trip that you were really looking forward to. What should you do?

LESSON 4

Chronic Disease and the Heart

You will learn . . .

- to identify some chronic diseases.
- what symptoms signal heart disease.
- how to reduce your risk of heart disease.

Vocabulary

- **chronic disease**, *D61*
- **heart disease**, *D62*
- **heart attack**, *D63*
- **risk factor**, *D64*

You know that getting physical activity can protect your heart. Heart disease is a serious illness. Healthful habits you start now help protect you from this disease for the rest of your life.

Chronic Disease

Most communicable diseases are acute diseases. *Acute diseases* are those that last only a short time. You usually recover from these diseases in a few days or weeks if you do what a doctor says.

Some communicable diseases and many noncommunicable diseases are chronic diseases. **Chronic** (KRAH•nik) **diseases** are conditions that last a long time or keep coming back.

Some chronic diseases are inherited diseases. An *inherited disease* is a disease caused by traits you get from your parents. *Cystic fibrosis* (SIS•tik•figh•BROH•suhs) is an inherited disease in which thick mucus clogs the lungs. People who have cystic fibrosis may have trouble breathing. *Sickle-cell anemia* (SIK•uhl•sel•uh•NEE•mee•uh) is another inherited disease. In this disease, the blood cells change shape. These cells don't move through blood vessels easily. This can cause pain and harm organs.

Other chronic diseases develop during a person's life. Pathogens cause some of these diseases. Bacteria cause Lyme disease, for example. Lyme disease can cause problems with the joints, heart, and nervous system. These problems can last throughout a person's life.

Sometimes substances around you can cause disease. Chemicals in the air or water can increase the risk of some kinds of cancer. You can reduce your risk of some chronic diseases. Practicing healthful habits can help. You will learn more about these habits in the next few lessons.

ACTIVITY

Science LINK

Explore Gene Therapy

Many chronic diseases are inherited. They are caused by problems with genes, which are tiny parts of cells that determine the traits each person inherits from his or her birth parents. Scientists are working to find ways to fix or replace genes that cause disease. Use the library or Internet to find out more about gene therapy. Make a poster or presentation for your class to share what you learned.

What is the difference between acute and chronic disease?

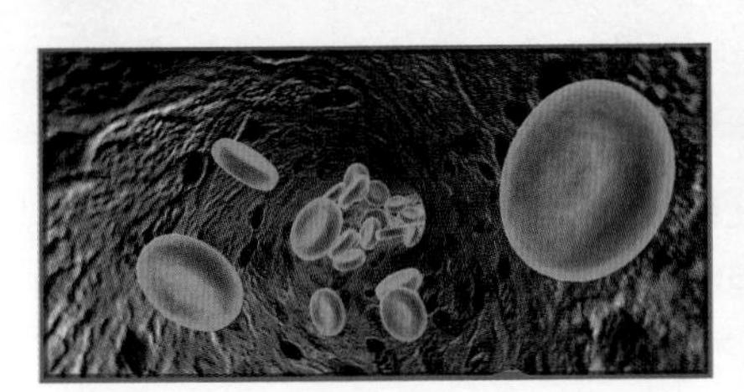

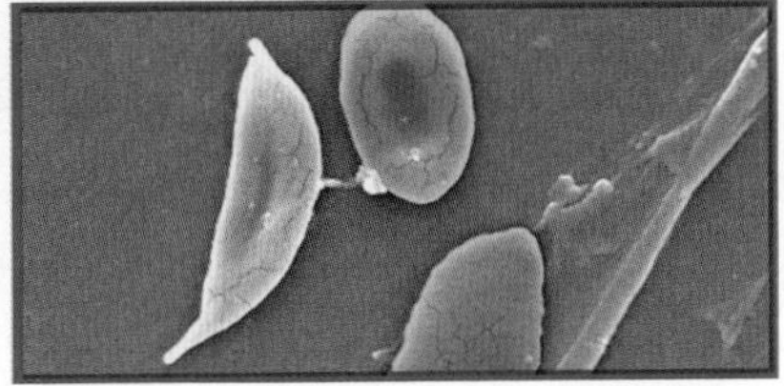

The photo on the left shows normal red blood cells. The photo on the right shows the cells of a person with sickle-cell anemia. The sickle-shaped cells cannot pass through blood vessels as easily as the normal cells.

Heart Disease and Strokes

Science LINK

Model a Clogged Artery

Pour sand through a paper towel tube. Notice how fast it moves through the tube. Now cut the tube open. Tape or glue cotton balls to the inside of the tube. Then roll the tube back up and tape it closed. Try pouring sand through it again. How does the cotton change how the sand moves? In coronary heart disease, fat and cholesterol clog arteries and make it harder for blood to flow.

Heart disease is a disease of the heart or the heart's blood vessels. It's the leading cause of death for men and women in the United States. About 950,000 people die from heart disease and strokes in the United States every year.

The most common type of heart disease is *coronary* (KAWR•uh•nayr•ee) *heart disease.* The arteries that bring blood to the heart become narrow. Usually this happens because fat and cholesterol build up in the arteries. You will learn more about coronary heart disease in the rest of this lesson.

There are many kinds of heart disease. *Congenital* (kuhn•JEN•uh•tuhl) *heart disease* is caused by a heart problem present at birth. Congenital heart disease can change how blood flows. For example, blood may not flow to the lungs. Many defects can be fixed with surgery.

Another kind of heart disease is *rheumatic* (roo•MAT•ik) *heart disease.* This disease is caused by strep throat that wasn't treated. It causes damage to certain areas of the heart. See a doctor right away if you have symptoms of strep throat.

A stroke is another kind of problem of the circulatory system. A *stroke* is a lack of oxygen to the brain due to a blocked or burst blood vessel. The risk of stroke increases if fat and cholesterol build up in arteries or veins in the head or neck.

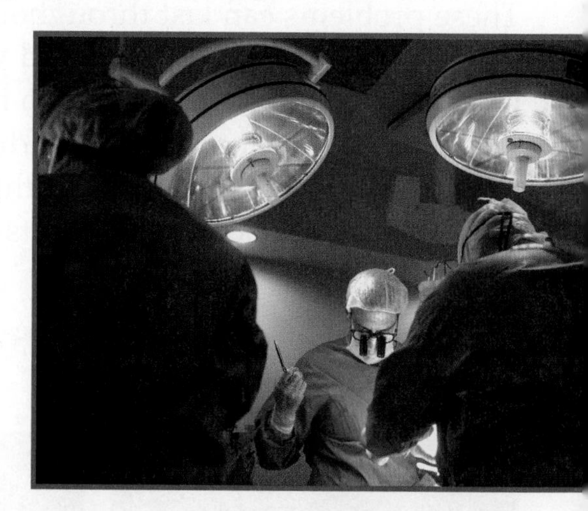

▲ **Doctors can repair many congenital heart problems. Sometimes they can do surgery on very young infants.**

Pixtal/AGE Fotostock

Coronary Heart Disease

The heart is a muscle. Like other muscles it needs oxygen. Blood brings that oxygen to the heart. The heart can't get enough blood if the coronary arteries are blocked. The lack of oxygen can cause pain. *Angina pectoris* (an•JIGH•nuh PEK•tuh•ruhs) is chest pain caused by a lack of oxygen in the heart. Angina occurs most often during exercise or when a person is stressed.

What if fat and cholesterol block an artery so much that blood can't get through? This can cause a heart attack. A **heart attack** is a sudden lack of oxygen to the heart due to a blocked blood vessel. If the heart does not get oxygen soon, it will not pump blood to the rest of the body. The rest of the body will not get oxygen. The person's brain and other organs will shut down.

A heart attack is an emergency. If a person having a heart attack does not get help quickly, he or she can die within a few minutes. The box lists symptoms of a heart attack. If you see someone having these symptoms, tell an adult right away. Call for medical help right away.

Heart Attack Symptoms

- **Chest pain, discomfort, or pressure**
- **Pain in the back, jaw, throat, or arm**
- **Discomfort above the stomach area**
- **Sweating, nausea, vomiting**
- **Difficulty breathing**
- **Dizziness and fainting**
- **Weakness and anxiety**

What is the most common type of heart disease?

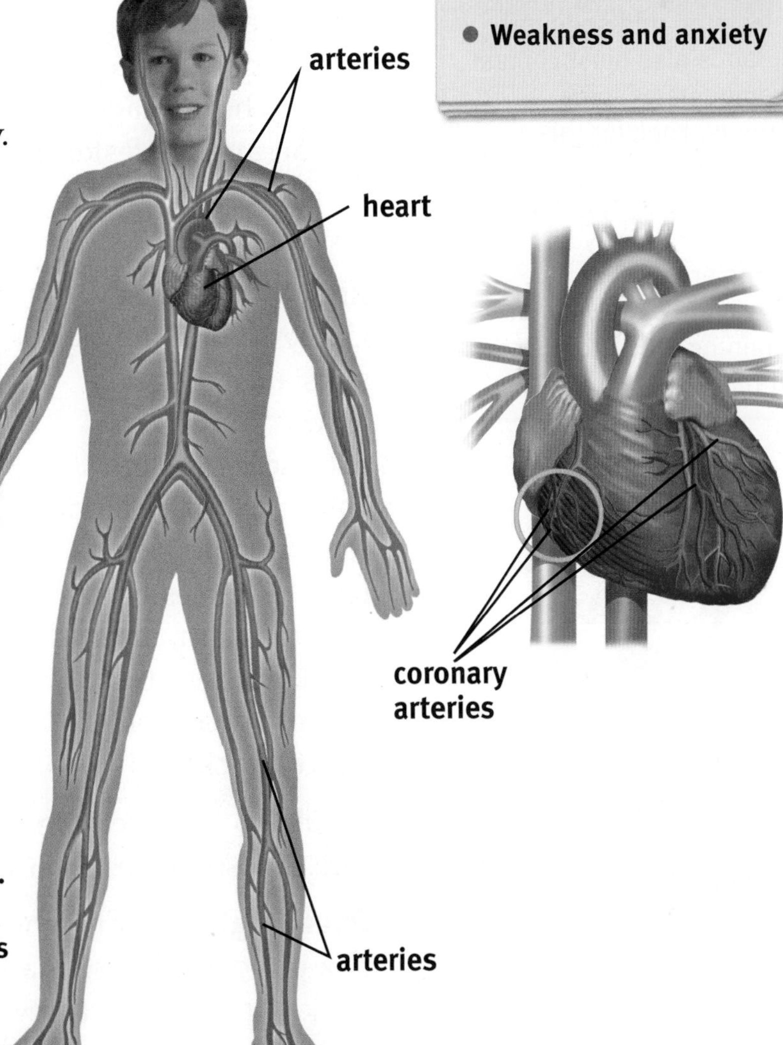

▶ **The coronary arteries bring blood to the heart. If these arteries are clogged, the person gets coronary heart disease.**

Risk Factors for Heart Disease

Not everyone has the same chance of getting heart disease. Some people have more risk factors than others. A **risk factor** is anything that increases the chance of loss or harm. Here are some risk factors for coronary heart disease.

Heredity The risk for heart disease can run in families. Suppose your grandfather had heart disease as a young man. You might have a higher risk of getting the disease. This doesn't necessarily mean you will get it, though.

High Blood Pressure Blood pushes on the walls of arteries as it flows. This causes blood pressure. If the pressure is too high, it can damage the artery walls. The damage lets fat and cholesterol attach to the arteries more easily.

▲ **Why is it important to have your blood pressure checked regularly?**

High Cholesterol Cholesterol is a waxy substance found in blood. Your body produces it. Some comes from the foods that you eat. If the level of cholesterol gets too high, it could block arteries and cause a stroke or heart attack.

Smoking The nicotine in tobacco speeds up the heart rate. It also makes arteries narrower.

Overweight People who are overweight have a higher risk of heart disease than those at a healthful weight. Lack of physical activity also increases the risk.

Dynamic Graphics Group/Creatas/Alamy

Check for "Heart-Smart" Foods

The American Heart Association puts its "heart-check" symbol on foods that meet its guidelines. These foods are low in fat, cholesterol, and sodium. The foods also must contain set amounts of some nutrients. Look for this symbol on food products the next time you visit a supermarket with your parents or guardian. Then create a daily menu containing food products that have the symbol.

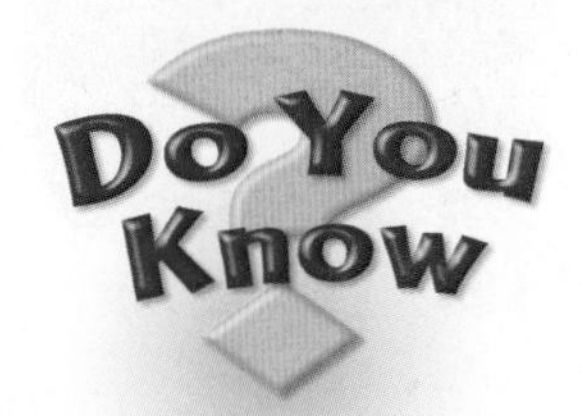

Myth Smokeless tobacco is safe to use.

Fact Smokeless tobacco is related to cancers of the mouth, esophagus and stomach, and pancreas. To reduce the risk of these diseases, the rule is the same for smoking: do not use tobacco.

Reduce Your Risk

You can practice habits now to reduce your risk of heart disease.

- **Stay at a healthful weight.** Your heart won't have to work as hard.
- **Limit the amount of fat and salt you eat.** Eating less fat helps keep fat from building up in your blood vessels. Too much salt can raise your blood pressure.
- **Get plenty of physical activity.** Physical activity makes your heart stronger. It also helps you stay at a healthful weight.
- **Manage stress.** Stress makes your heart work harder. Talk with an adult if you feel stress.
- **Do not smoke.** Avoid secondhand smoke, too.

How does physical activity help reduce your risk of heart disease?

ACTIVITY
LIFE SKILLS
CRITICAL THINKING

Set Health Goals

Darlene's father and grandfather both had heart disease before they were 50 years old. She has learned that exercise can help protect her heart. She wants to set a goal to get more exercise. Write a plan Darlene could use.

1. **Write the health goal you want to set:** I will get physical activity every day.
2. **Explain how your goal might affect your health.** Write a summary that explains how physical activity will help Darlene's heart. Tell how her family history affects her risk of heart disease.
3. **Describe a plan you will follow to reach your goal.** Make a Health Behavior Contract for Darlene. Include 30 minutes a day of activity in your plan. Tell from whom she may get help—a parent, a guardian or a coach.
4. **Evaluate how your plan worked.** Suppose Darlene comes to you at the end of a week. How could she check her progress? What questions would you ask her? Write your questions on the back of the contract.

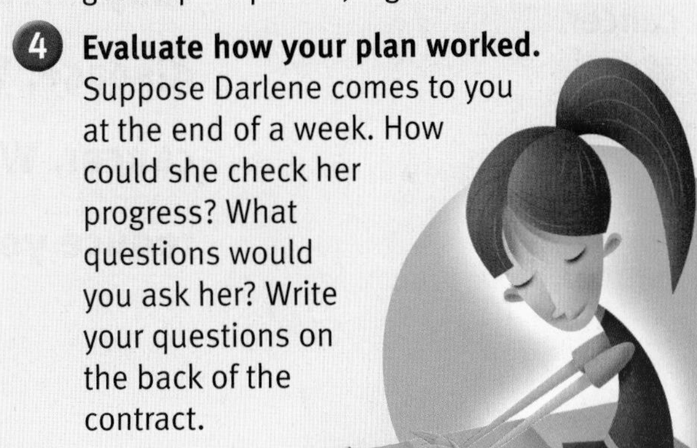

LESSON REVIEW

Review Concepts

1. **Define** a chronic disease.
2. **List** the symptoms of a heart attack.
3. **Explain** the major risk factors for coronary heart disease.

Critical Thinking

4. **Contrast** Explain the difference between congenital, rheumatic, and coronary heart disease.
5. LIFE SKILLS **Set Health Goals** Choose a behavior from the list on this page. Explain how it can protect your heart.

LESSON 5

Chronic Disease: Cancer

You will learn . . .

- how cancer effects the body.
- ways to reduce the risk of and treat cancer.

Vocabulary

- cancer, *D67*

About one million people every year find out that they have cancer. Cancer is a serious chronic disease. You can help protect yourself from cancer. Wearing sunscreen is one way you can reduce your risk of skin cancer.

What Is Cancer?

Cancer is a disease in which cells multiply in ways that are *abnormal,* or not normal. A group of these abnormal cells is called a *tumor* (TOO•muhr). Some tumors are *benign* (buh•NIGN), or harmless. Others are *malignant* (muh•LIG•nunt), or cancerous. Usually cancer starts in one body part. Cancer cells can then spread and destroy normal cells around them. Cancer can spread to other body parts. It can seem to go away, but then come back. Many kinds of cancer can be treated with medication or surgery.

In men the most common type of cancer is prostate (PRAHS•tayt) cancer. The prostate gland is part of the male reproductive system. In women the most common type of cancer is breast cancer. For both men and women, lung cancer is the next most common. Then comes cancer of the colon and rectum. The colon and rectum are parts of the large intestine.

What is cancer?

ACTIVITY

Math LINK

Analyze Graphs

Read the graphs below. Answer these questions.

1. For women, about what fraction of cancer cases are breast cancer?
2. Look at the graph for men. Prostate, lung, and colon and rectum cancer are the most common types of cancer. Are more or less than half of the cases of cancer one of these three types?
3. Write another problem that could be answered using the graphs. Trade with a partner and answer each other's question.

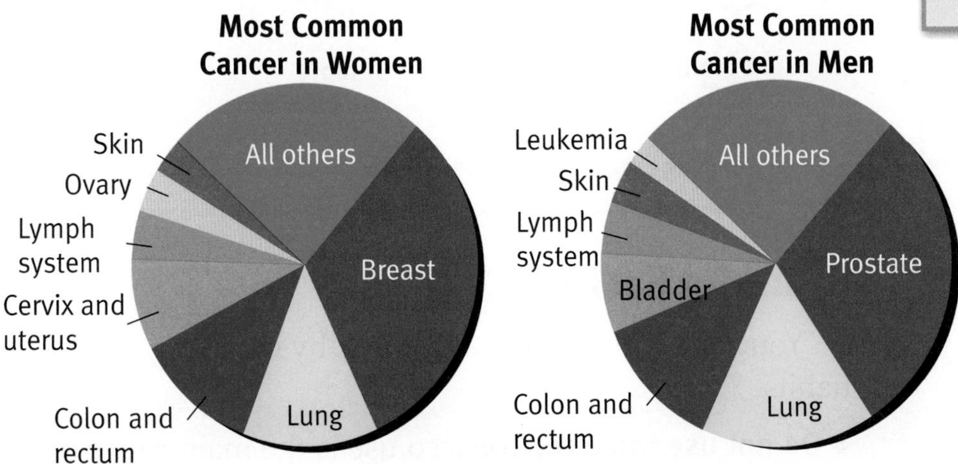

▲ **These graphs show the most common types of cancer in men and women. The cervix (SUHR•viks) and uterus (YOO•tuh•ruhs) are parts of the female reproductive system. The lymph (LIMF) system carries proteins, water, and white blood cells to the body. Leukemia (loo•KEE•mee•uh) is a cancer of the blood and the organs that form blood.**

Reducing the Risk

Heredity increases the risk of some kinds of cancer. There are also ways to reduce the risk.

Warning Signs

Many types of cancer can be treated if they are discovered early. The word CAUTION can help you remember cancer warning signs. See a doctor right away if you see any of these signs.

Healthful Habits

You can reduce your cancer risk by choosing healthful habits.

- **Do not use tobacco.** Tobacco use is the main cause of cancers of the lungs, mouth, and throat. Women who smoked as teens may have an increased risk of breast cancer. Secondhand smoke from tobacco can increase the risk of lung cancer, too.

MAKE a Difference

Gifts for Families

Nicholas Marriam of Edgewater, Maryland, had cancer. He spent 172 days being treated at a children's hospital. He saw that his parents became tired and stressed as they cared for him. When he went home, Nicholas wanted to help the families of people with cancer find ways to take care of their mental and emotional health. He began collecting movie tickets, gift certificates, health care products, and coupons donated by local businesses. He created gift bags with these items. He delivered the bags to children and their families at the hospital. How do you think Nicholas' own experience inspired him to help others?

Myth Some people do not need to use sunscreen because they do not sunburn.

Fact Everyone should wear sunscreen with an SPF of at least 15. Sunscreen reduces the risk of skin cancer for all skin types.

- **Protect your skin from the sun.** The sun's harmful rays increase the risk of skin cancer. You can reduce your risk. Avoid the sun between 10 A.M. and 4 P.M. Use sunscreen with an SPF of at least 15. Don't use tanning lamps or booths.
- **Check your skin regularly for signs of skin cancer.** Watch for any change in your skin or in a mole. Also watch for any sores that don't heal.
- **Eat healthfully.** Eat plenty of whole grains, fruits, and vegetables. The fiber in these foods helps reduce the risk of colon cancer. The vitamins in fruits and vegetables can also help reduce your risk of cancer. Avoid fatty foods. This may reduce the risk of some kinds of cancer.

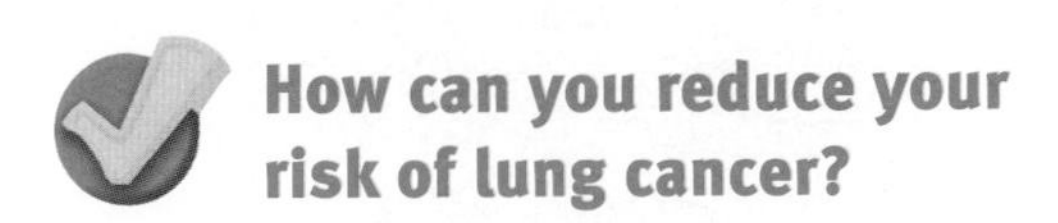

How can you reduce your risk of lung cancer?

CRITICAL THINKING

Be a Health Advocate

Make reminder cards to help your family and friends remember to reduce their risk of skin cancer.

1. **Choose a healthful action to communicate.** You know that protecting yourself from the sun reduces your risk of skin cancer.
2. **Collect information about the action.** Work with a classmate to identify three ways to protect yourself from the sun.
3. **Decide how to communicate this information.** Design a card that gives tips for protecting yourself from the sun.
4. **Communicate your message to others.** Make copies of your card. Give the card to others to remind them how to protect themselves.

Be Safe in the Sun

1. Wear sunscreen.
2. Wear sunglasses.
3. Wear a hat.

SPF 15

LESSON REVIEW

Review Concepts

1. **Describe** how cancer affects the body.
2. **List** the seven warning signs of cancer.

Critical Thinking

3. **Infer** How can you tell that cancer is a chronic disease, not an acute one?
4. LIFE SKILLS **Practice Healthful Behaviors** How can eating healthful foods reduce your risk of cancer?
5. LIFE SKILLS **Be a Health Advocate** Design an advertising campaign to teach people the warning signs of cancer. Write a paragraph describing your campaign.

Learning LIFE SKILLS

CRITICAL THINKING

Manage Stress

Problem Jason's grandmother just found out she has cancer. Jason is having trouble paying attention in school. He isn't sleeping well. His stomach hurts. What could be wrong?

Solution When a family member has a chronic illness, it can cause stress for everyone else in the family. There are ways Jason can manage his stress.

Learn This Life Skill

Follow these steps to help you manage stress. The Foldables™ can help.

1 Identify the signs of stress.

Jason's stomachache is a sign of stress. So is his trouble sleeping. Stress can make it hard to pay attention in school. What are some other signs of stress?

2 Identify the cause of stress.

A family member's illness can cause a lot of stress. Why might Jason feel stress when his grandmother is ill?

3 Do something about the cause of stress.

What can Jason do to help feel less stress? He can talk to his parents. He can visit his grandmother. What else can he do?

4 Take action to reduce the harmful effects of stress.

Jason can protect his body from the stress he feels. He can listen to relaxing music before he goes to bed. This may help him sleep. He can make sure he follows a healthful diet. What else can he do?

Identify the signs of stress. ①
Identify the cause of your stress. ②
Do something about the cause of your stess. ③
Take action to reduce the harmful effects of stress. ④
Manage Stress

Practice This Life Skill

With a group, make a booklet for people who have ill family members. Explain how these people can use the four steps to help manage stress.

LESSON 6

Other Chronic Diseases

You will learn . . .

- what diabetes is and how to manage it.
- what asthma and allergies are and how to manage them.
- what epilepsy and arthritis are and how to manage them.

Many people live with chronic diseases. They can learn how to manage these diseases. They can use this knowledge to have active, healthy lives.

Diabetes

You have read that diabetes occurs when there is too much sugar in a person's blood. Insulin is a hormone that helps body cells use sugar. The body cells of a person with diabetes can't get enough energy from sugar.

There are two main forms of diabetes. Type 1 diabetes begins in childhood. People with this form of the disease produce little or no insulin. They must take shots of insulin every day. They also eat foods low in sugar.

A person with type 2 diabetes does make some insulin. The person's body does not use the insulin well. Most people with this kind of diabetes are adults. Young people can also develop it. People with type 2 diabetes can often control it with diet and exercise. They sometimes need medicine if diet and exercise do not help enough.

The number of people with diabetes has increased in recent years. Find out more about the risk factors for type 1 and type 2 diabetes. Report what you learn using the e-Journal writing tool.

Managing Diabetes

People with diabetes have an increased risk for other health problems. These include blindness, heart disease, and kidney disease. They can reduce their risk by managing their diabetes.

People who have diabetes must have blood tests regularly. The blood tests tell how much sugar is in the blood. The people may then take insulin to keep their blood sugar at a healthful level. It's also very important for people with diabetes to maintain a healthful weight. They may need to follow a special diet.

Preventing Diabetes

You can help reduce your risk of type 2 diabetes. One of the most important things you can do is to maintain a healthful weight. People who are overweight are more likely to develop type 2 diabetes. Follow a balanced diet and get plenty of physical activity. Get regular medical checkups. The sooner a chronic disease is found and treated, the more successfully it can be managed.

How can you reduce the risk of type 2 diabetes?

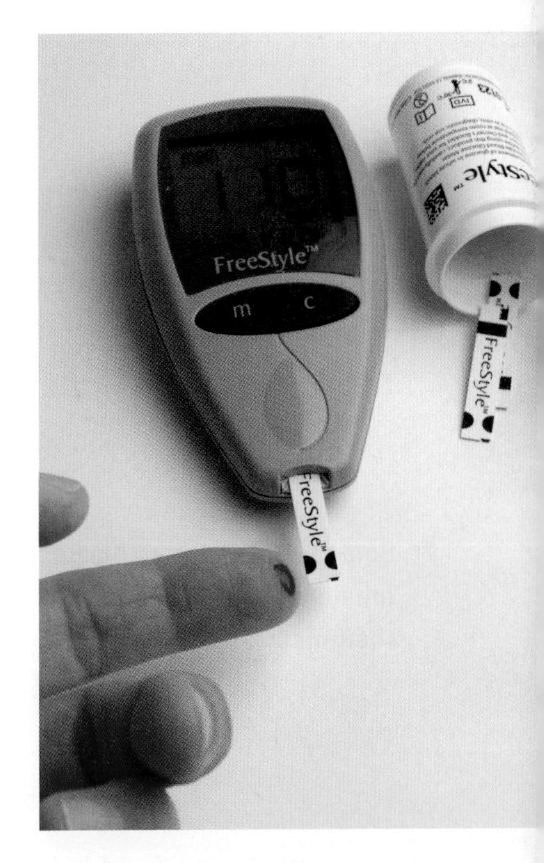

▲ **People with diabetes must have frequent blood tests to check the sugar levels in their blood.**

Asthma and Allergies

Asthma and allergies are two common chronic conditions that affect the lungs.

Asthma

Asthma (AZ•muh) is a chronic condition in which the small airways in the lungs become narrow. This makes it hard to breathe. A person having an asthma attack may wheeze and gasp for air. He or she may have a dry cough. His or her chest may feel tight. Get medical help right away if the person has trouble breathing. A school nurse can often help a person who is having an asthma attack. The person may need treatment in a hospital.

Substances or conditions that bring about asthma attacks are called *triggers.* Animal hair, dust mites, mold, plant pollens, tobacco smoke, and some medicines can all be triggers. Cold air, vigorous exercise, and stress can also cause asthma attacks.

Avoiding asthma triggers is the first step in managing the disease. A doctor may also prescribe medicine. Often people with asthma use an inhaler. An *inhaler* is a device that lets the person breathe medicine directly into the lungs.

Write About It!

Write an Explanation
Explain what a person should do when having an asthma attack. Include the symptoms of an attack and describe how the person could get help when at home, school, or at a park or playground.

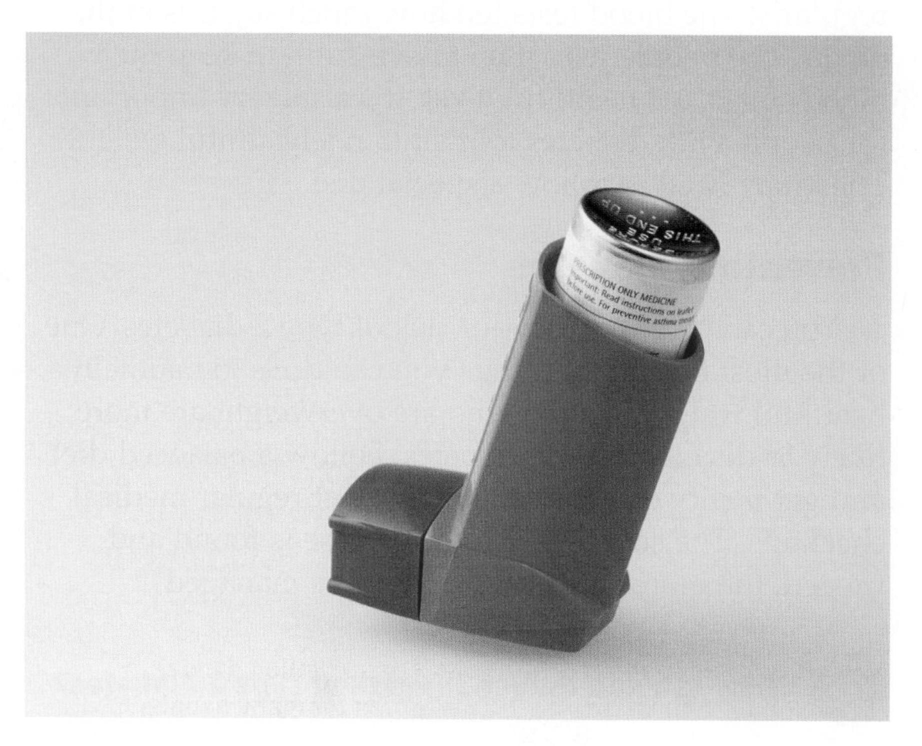

Some people with asthma use inhalers like this one to take their medication.

Brand X Pictures/Alamy

Allergies

An *allergy* (AL•er•jee) is a reaction of the body to certain substances. The thing that causes an allergy is called an *allergen.* The immune system treats the allergen like a pathogen.

Almost anything can be an allergen. Animals, pollen, mold, and dust are common allergens. Many people are also allergic to foods, such as milk, eggs, nuts, or shellfish.

Allergy symptoms vary. Often people sneeze and have a runny nose. Their eyes get itchy and watery. Some people may get headaches or skin rashes. People with severe allergies may have trouble breathing. A person who has a severe allergic reaction needs medical help right away. People who know that they have severe allergies may carry medicine to treat a reaction.

Sometimes a person can manage an allergy just by avoiding the allergen. In other cases an allergist may recommend treating the allergy with medication. An *allergist* is a doctor who treats people with allergies. The allergist may give a person shots. The shots make the person less sensitive to an allergen. Sometimes the shots can keep the person from reacting to the allergen at all.

Write a Commercial

Locate ten food labels that include one of the allergens listed on this page. Work with a partner to create a television commercial to warn people about the allergens in the foods.

List five common allergens.

▲ **Cats and other animals have a substance called dander in their hair. Animal dander is an allergen for many people.**

Epilepsy and Arthritis

There are many other chronic diseases. You read about cystic fibrosis and sickle-cell anemia in Lesson 4. Two other common chronic diseases are epilepsy and arthritis.

ACTIVITY

Science LINK

Explore Arthritis

In arthritis the padding between bones can wear away. Take two pieces of wood. Try to slide one along the other. This represents the bones of a person with arthritis. Then wrap one piece in aluminum foil or plastic wrap to represent the padding. Slide the wood again. Which was easier to move?

Epilepsy

In *epilepsy* (EP•uh•lep•see) nerve messages in the brain are disturbed for brief periods of time. A person with epilepsy has seizures. In a *seizure* the electrical signals in the brain are disrupted. The person loses control of his or her thinking and movement for a short time.

Sometimes an injury to the brain causes epilepsy. Heredity and some kinds of infections can also cause epilepsy. In many cases nobody knows what causes it.

There are several kinds of seizures. Some seizures may last only a few seconds. A person may just blink rapidly or seem not to be paying attention. Some seizures may last for several minutes. During these seizures a person's muscles may jerk. The person may fall to the ground. He or she may even pass out.

Many people with epilepsy can control their seizures with medicine. The medicine helps reduce the number of seizures. It can sometimes stop the seizures completely. Some people need surgery.

If you see someone having a seizure, watch the person. Make sure that he or she isn't injured. Move objects away from the person so that he or she doesn't hit them. Don't try to stop the person from moving. Let the person rest after the seizure.

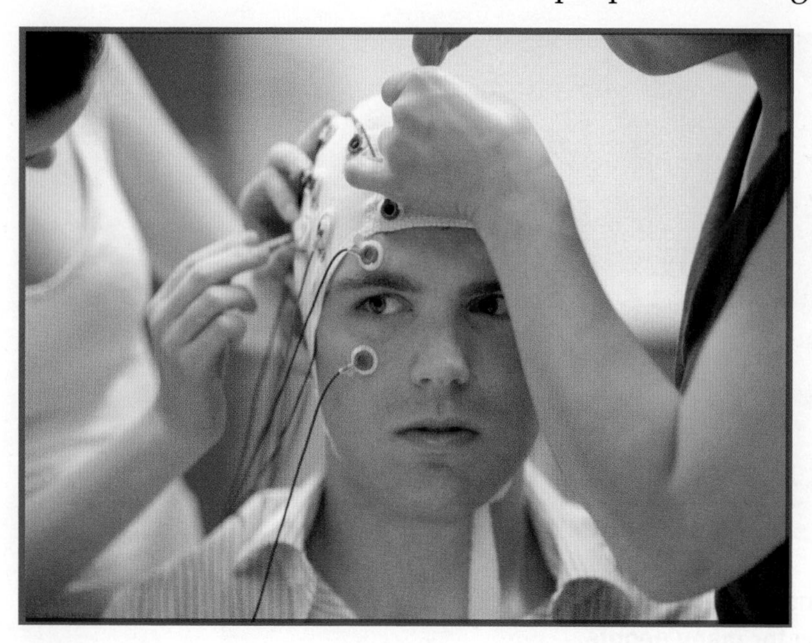

Scientists study the electrical signals in the brains of people with epilepsy. They are working to find ways to treat this disease.

Arthritis

Arthritis (ahr•THRIGH•tis) is the painful swelling of joints in the body. In one kind of arthritis, the padding between bones wears away. Another kind is a disease of the linings of the joints. Sometimes the body's immune system attacks healthy body parts. This can also cause arthritis. Injuries to joints can cause some kinds of arthritis. Being overweight can make arthritis worse.

People with arthritis can take medicines to ease the pain. Some of these medicines also reduce the *inflammation,* or swelling, of the joints. Exercise helps strengthen the joints. In severe cases a person may need surgery. The surgery can repair or replace a joint. Hip and knee joints are two joints that surgeons can replace. These treatments help many people with arthritis live active lives.

ACTIVITY

LIFE SKILLS

CRITICAL THINKING

Analyze What Influences Your Health

Knowing what influences your health is especially important if you have a chronic disease. Make a list of "Dos and Don'ts" for people with asthma.

1. **Identify people and things that can influence your health.** List as many factors as you can that can influence the health of a person with asthma. For example, avoiding triggers could affect the person's health.
2. **Evaluate how these people and things can affect your health.** Describe how each of your choices affects the person's health.
3. **Choose healthful influences.** List the healthful influences in a column titled "Do."
4. **Protect yourself against harmful influences.** List the harmful influences in a column titled "Don't."

Make your list into a colorful brochure. Share your brochure with your class.

How does exercise help manage arthritis?

LESSON REVIEW

Facts and Skills

1. **Distinguish** between type 1 and type 2 diabetes and tell how each is managed.
2. **Describe** how the body responds to asthma and allergies.
3. **Explain** the characteristics of epilepsy and arthritis.

Critical Thinking

4. **Analyze** You are allergic to chocolate. A friend offers to share some trail mix with you. What should you do before you eat the trail mix?
5. LIFE SKILLS **Analyze What Influences Your Health** Suppose you have diabetes. What are some harmful influences you should protect yourself against?

CHAPTER 8 REVIEW

Use Vocabulary

- **antibiotic,** *D58*
- **antibody,** *D53*
- **cancer,** *D67*
- **heart attack,** *D63*
- **immunization,** *D53*
- **symptom,** *D57*
- **virus,** *D46*

Choose the correct term from the list to complete each sentence.

1. A drug used to kill or slow the growth of germs that cause disease is a(n) __?__.
2. A sudden lack of oxygen to the heart results in a(n) __?__.
3. A change in behavior or body function is a(n) __?__ of disease.
4. Something that keeps you from getting a disease is __?__.
5. A tiny particle that makes copies of itself inside a living cell is a(n) __?__.
6. A substance in your blood that helps fight pathogens is a(n) __?__.
7. A disease in which cells multiply in ways that are abnormal is __?__.

Review Concepts

Answer each question in complete sentences.

8. List the four most common pathogens that cause communicable disease.
9. How do your white blood cells protect you from disease?
10. What are four symptoms of flu?
11. What is type 1 diabetes?
12. What happens when a person has an allergy?
13. List the warning signs of cancer.
14. Explain what epilepsy is and how it can be managed.

Reading Comprehension

Answer each question in complete sentences.

Suppose that you get chicken pox. While you are ill, your body starts making antibodies. The antibodies attack the virus that causes chicken pox. These antibodies stay in your blood for a long time. They may even stay for the rest of your life. What happens if the virus that causes chicken pox gets into your body again? The antibodies destroy the virus.

15. When does your body make antibodies?
16. What happens if the chicken pox virus gets into your body again?
17. How else can you develop immmunization to a disease?

Critical Thinking/Problem Solving

Answer each question in complete sentences.

Analyze Concepts

18. Why is it important to know the warning signs of cancer and heart attacks?

19. Explain how maintaining a healthful weight can help reduce your risk of chronic diseases.

20. Summarize the stages of disease.

21. Why should you wash your hands after you blow your nose?

22. Why is it important for people with chronic diseases to learn to manage those diseases?

23. Suppose you fall and scrape your knee. Why should you both clean and bandage the scrape?

Practice Life Skills

24. **Manage Stress** You have a big test at school tomorrow. It's causing you stress. How can you protect your heart from the effects of stress?

25. **Make Responsible Decisions** You have type 1 diabetes. A friend is waiting for you to come out and play, but you haven't checked your blood sugar yet. Your friend says, "You don't need to do that. We aren't going to be eating anything." What should you do? Explain why your decision is responsible.

Read Graphics

Use the graph to answer the questions.

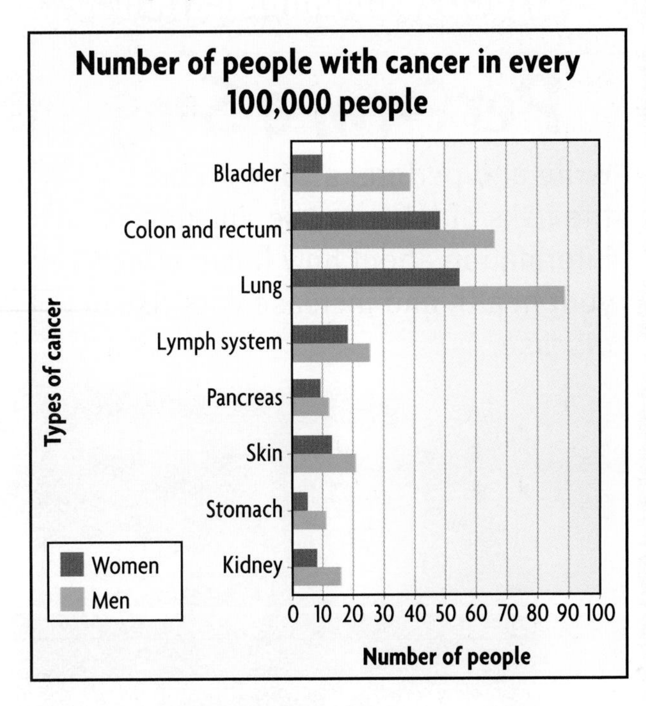

26. Overall, is the rate of these types of cancer greater for men or for women?

27. About half as many women as men get kidney cancer. For which other types of cancer is this true?

28. Suppose that the number of people who smoke tobacco goes down. How would you expect the graph to change?

ACTIVITIES AND PROJECTS

Effective Communication

Perform a Song

Write and perform a song about the risks of tobacco use. Include information about how it can affect your health and increase your risk of disease.

Self-Directed Learning

Draw a Diagram

Choose three drugs that were discussed in this unit. Find out more information on how each one affects the body. Draw a diagram to summarize what you learn.

Critical Thinking and Problem Solving

Make a Plan

Choose either heart disease or a type of cancer. Find out what you can do to reduce your risk of the disease. Write a step-by-step plan for new habits you can practice.

Responsible Citizenship

Create an Announcement

Write a script for a television announcement to reduce the risk of diseases carried by insects. Describe the images you would show in the announcement.

UNIT E

Community and Environmental Health

CHAPTER 9

Consumer and Community Health, *E2*

CHAPTER 10

Environmental Health, *E30*

CHAPTER 9

Consumer and Community Health

Lesson 1 • What Smart Consumers Know E4

Lesson 2 • Help for Consumers E10

Lesson 3 • Planning a Healthful Community E16

Lesson 4 • Careers in Health and Fitness . . . E22

Ken Karp/McGraw-Hill Education

What Do You Know?

This chapter tells you how to be a wise consumer. What do you know about being a consumer? Decide whether these statements are **true** or **false**.

Companies use ads to persuade you to buy their products.
A consumer reporter can help if you have a problem with a product.
Many factors influence you to buy a product.
You can believe all the information you get from the Internet.
There is nothing you can do if a product you buy does not work.
Ads tell you all you need to know about a product.

The first three statements are true, and the last three are false. You will learn how to choose health products and services in **Consumer and Community Health**. You'll also learn about health in your community and health careers.

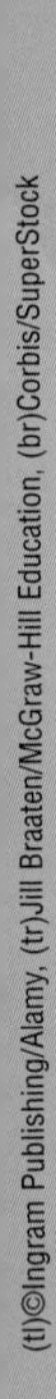

LESSON 1

What Smart Consumers Know

You will learn . . .

- what health-care products and services are.
- factors that can influence choices about health-care products and services.
- how technology affects health and health care.

Every day you make choices about what you buy and use. What is behind the choices you make? Getting the right information can help you make responsible choices.

Vocabulary

- **health-care product**, *E5*
- **health-care service**, *E5*
- **consumer**, *E5*
- **media**, *E6*
- **advertising**, *E6*
- **appeal**, *E6*
- **generic**, *E7*

Health-Care Products and Services

A **health-care product** is an item used for physical, mental, or social health. Toothpaste, deodorant, and vitamins are all examples of health-care products. A **health-care service** is something another person does that improves your physical, mental, or social health. Doctors and dentists provide health-care services. So do hospitals and fitness centers.

A **consumer** is a person who judges information and buys and uses products and services. You are a consumer. You need health facts to make smart choices. How do you know which health-care products and services are right for you? You can use your health knowledge to decide.

What is a consumer?

CAREERS

Consumer Reporter

Consumer reporters help people who have problems with products or services. They investigate complaints from consumers. They do television reports or write articles.

If you'd like to be a consumer reporter, study English and writing. It helps to be interested in current events. Consumer reporters also need to like working with others.

Reading a Product Label

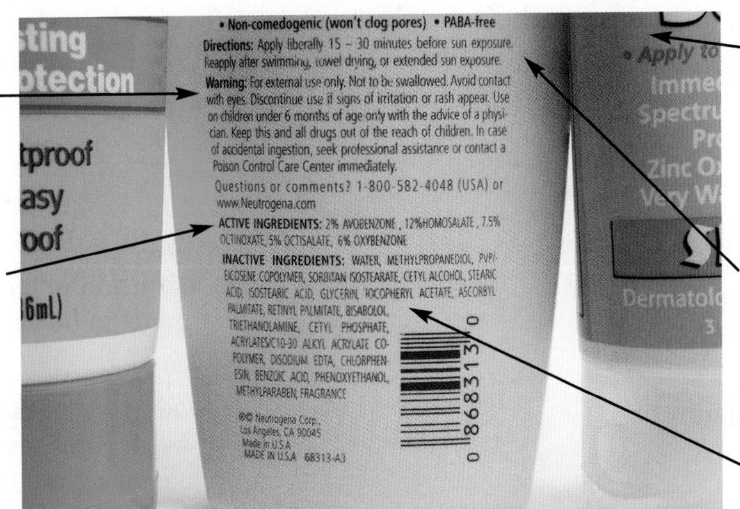

Warnings Lists any safety warnings about the product.

Active Ingredients Lists the ingredients that make the product work.

Uses Tells what the product is used for. This product blocks out the sun's UV rays.

Directions Tells how to use the product.

Inactive Ingredients Lists the other ingredients in the product.

Choosing Wisely

Write About It!

Write a Persuasive Commercial Choose a health-care product. With a partner, write a TV or radio commercial to advertise your product. Your ad should focus on how the product can improve people's health.

What makes you choose a particular health-care product? Maybe your family or friends use it. Maybe the price is low. Many factors can affect the choices that you make.

Media Influences

The media can influence your choices. The **media** are sources of news and information. They include TV, radio, magazines, and newspapers. The media often carry advertising. **Advertising** is a form of selling products and services. Advertisements are also called *ads*.

Ads often make appeals. An **appeal** is a statement used to get consumers to buy a product or service. Some ads show popular actors or sports stars. The company hopes you will buy the product because you like the person. Other ads show young people using a product. They show the people having fun with many friends. The company wants you to think that the product will make you happy and popular.

Ads often leave out information. Companies don't want you to know things that may make you choose not to buy their products. A fast-food ad might not tell you about the fat content of the food. Use your health knowledge to fill in the information that an ad doesn't give you.

Look for a seal of approval on a product. This means that the product meets standards set by an organization.

Seals of Approval

Organization	What It means
American Dental Association	The American Dental Association judges the product to be safe and effective.
American Heart Association	The food in the package meets the American Heart Association's standards for saturated fat and cholesterol for healthy people over age 2.
USDA Organic	The food in the package meets the U.S. Department of Agriculture's standards for organically grown food.

How can you make smart decisions on products and services? First ask yourself, "Do I really need this?" If you do, then learn about the product. There are many ways that you can get reliable health information. You can talk to a doctor, pharmacist, teacher, or your parents or guardian for advice. You can get information from a library and from government groups. You can also use technology, such as the Internet and television. You'll learn more about these sources on the next page.

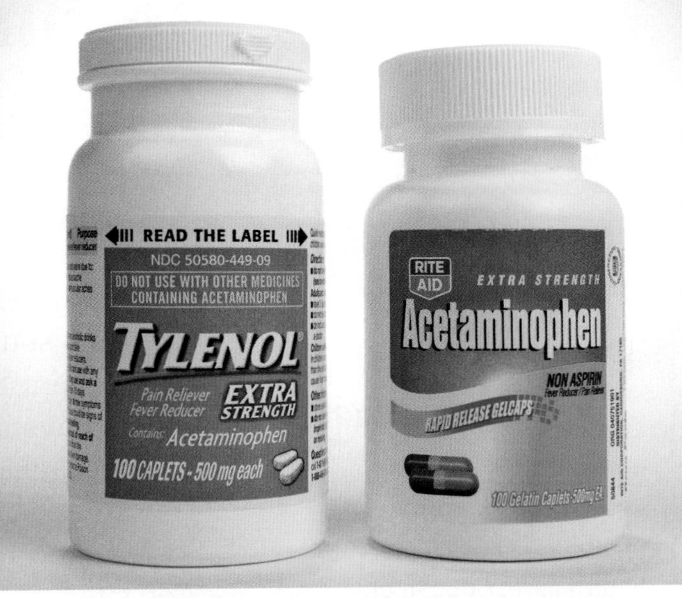

▲ Generic products, such as the one on the right, contain the same ingredients as brand-name products.

What is the purpose of advertising?

ACTIVITY
LIFE SKILLS
CRITICAL THINKING

Make Responsible Decisions

You are comparing two skin lotions. One is a well-known brand of lotion. The other is a generic lotion. **Generic** products do not have a brand name, though they usually work as well. They often cost less. Your family is trying to save money. Your friends use the well-known brand. Which one should you use? Use the steps for making a responsible decision.

1. **Identify your choices. Check them out with your parent or another trusted adult.** You can buy either lotion.
2. **Evaluate your choices. Use the *Guidelines for Making Responsible Decisions*™.** Ask yourself the six questions for each choice.
3. **Identify the responsible decision. Check this out with your parent or another trusted adult.** Which is the better choice for you and your family?
4. **Evaluate your decision.** Why is it responsible to buy the lotion you chose? Explain your decision to a classmate.

***Guidelines for Making Responsible Decisions*™**

- **Is it healthful?**
- **Is it safe?**
- **Does it follow rules and laws?**
- **Does it show respect for myself and others?**
- **Does it follow family guidelines?**
- **Does it show good character?**

Technology and Health Care

ACTIVITY

Science LINK

Present Health-care Technology

With a small group, brainstorm ways that technology has changed health care. Choose one of the topics that interests you. You might choose the Internet. You might choose new tools that doctors use to see inside the body. Find out more about the technology you chose. Make a presentation to show your class how that technology has changed health care.

The Internet can help you find health information fast. But it can also give poor advice or false information. Be sure that the facts you find come from reliable sources. One such source is the U.S. government. Many government Web sites have information about health. Others have information about health-care products.

Other reliable sources are trusted health organizations. The American Heart Association, the American Cancer Society, and the American Lung Association all have Web sites.

At some Web sites you can type in questions that a doctor or nurse will answer. Check with your parents or guardian before you use the Internet. They can help you find reliable sources. Always see your doctor about serious health concerns.

▼ **Ask your parent or guardian to show you some reliable Internet sources of information.**

TV Guidelines

How much TV do you watch? How do you choose what programs to watch? It's wise to choose TV programs carefully. It's also a good idea to limit the hours you watch TV.

Many TV programs show violence. Watching violence on TV may make young viewers fearful. It may also make them more likely to behave in harmful ways. Watching TV takes time you could spend with other people. It uses up time that you might spend being physically active.

TV programs often show many ads. These ads may make you want things that you do not need. The ads also sometimes show behaviors that are not healthful. These ads can influence people to make choices that are not responsible. If you limit the time you watch TV, you will reduce the number of ads you see. This reduces the chance that you will be influenced in harmful ways.

Tips for TV Watching

- Plan with your parents or guardian which shows you will watch.
- Finish your homework before you watch TV.
- Make a list of other activities that you can do instead of watching TV. Then do them!

What are two reliable sources of health-care information on the Internet?

LESSON REVIEW

Review Concepts

1. **Define** *health-care product* and *health-care service*. Give an example of each.
2. **Identify** what you should do before choosing a health-care product or service.
3. **Explain** the positive and negative effects technology can have on your health.

Critical Thinking

4. **Analyze** How could you evaluate the claims an ad makes about a product?
5. LIFE SKILLS **Practice Healthful Behaviors** You are inside on a rainy Saturday. You don't want to spend too much time watching TV. What other healthful behaviors can you practice?
6. LIFE SKILLS **Make Responsible Decisions** How can asking yourself, "Do I really need this?" help you use the *Guidelines for Making Responsible Decisions*™?

LESSON 2

Help for Consumers

You will learn . . .

- how the government and private groups help protect consumers.
- strategies to help manage time and money.

Vocabulary

- **budget**, *E12*
- **schedule**, *E13*

Has your family ever bought a product that didn't work properly? Often you can return the product to the store and get your money back. If that doesn't work, there are people and groups who can help you.

©Hero/age fotostock

Protecting Consumers

Government and private groups work to make sure that products do what they should. These groups also help consumers when products do not work correctly.

Government

Three U.S. government agencies help protect consumers. The Food and Drug Administration (FDA) makes sure that food, drugs, and medical devices are safe. The FDA also makes rules about how companies must label these products.

The Federal Trade Commission (FTC) makes rules about how companies can do business. The FTC makes sure that ads tell the truth about a product. You can get information from the FTC about how to choose safe products and services.

The Consumer Product Safety Commission (CPSC) warns consumers about products that are not safe. The CPSC may announce a recall. A *recall* is a warning to consumers that a product is not safe. For example, the CPSC may announce that one brand of battery could leak. Consumers who have those batteries should take them back to a store. They can get a new product that is not damaged, or they can get their money back.

Private Groups

A *consumer advocacy group* helps consumers when they have problems with products or services. Often a group can do more than an individual. One person would have a hard time making a company change what it does. A company is more likely to listen to a group of people. This is because they do not want that group of people to stop using the company's products or services. Consumer advocacy groups bring many people together to get companies to fix problems.

What does a consumer advocacy group do?

ACTIVITY

Con$umer Wi$e

Design a Brochure

In the brochure, tell people about the Better Business Bureau (BBB) in your community. The BBB is a private consumer advocacy group. The BBB encourages businesses to be honest and fair to consumers. It also warns consumers about companies that are not fair or honest. You can complain to the BBB if a company has not been honest or fair to you.

Budgets and Schedules

Budget Benefits

Budgeting helps you plan how to save and spend your money. Learning how to budget now will help you as you get older and earn more income. Research the benefits of budgeting. Use the e-Journal writing tool to summarize the information you found.

You can choose wisely when you buy health products and services. You can also plan how to spend your time and money.

Plan Your Money

A **budget** is a plan for spending and saving money. A budget will help you have enough money for the things you need. Follow these steps to plan a budget.

List your income. *Income* is the amount of money you expect to receive in a certain period of time. This amount might include your allowance and any money you earn.

List your expenses. *Expenses* are items you spend money on each week. They may include your lunch money or bus fare, for example. Add up all your expenses.

Plan to save money. Your expenses should be less than or equal to your income. If there is money left over, plan to save some of it. Maybe you want a new pair of shoes. If you save a little every week, you can save enough money to buy the shoes.

Plan for things you want. If you have money left over, you might plan to use some of it on something you want. Maybe you want to go to the movies. Include the price of the ticket in your budget. That way you can be sure you have enough money to go.

▼ **Computer programs can help you make budgets. If this were your weekly budget, how much would be left to spend on things you want?**

My Budget

Income	Savings	Things I need	Things I want
$10.00 from Aunt Julie	$3.00	$5.50 new markers for art class	
$5.00 allowance		$5.00 museum ticket for field trip	

Comparison shop. To get more for your money, comparison shop. *Comparison shopping* is checking the price of an item in several stores or comparing similar products in the same store before you buy. Two products may have different prices even though they have the same ingredients.

Plan Your Time

A schedule can help you manage your time. A **schedule** is a list of times, events, and activities to do. It can help you make time for the activities you need and want to do. Follow these steps to make a schedule.

- **Write down what you must do.** Start with events that have set times. These might include school, sports practice, or a family party. Then write down when you'll do your homework and your chores.
- **Add what you want to do.** Include time for exercise and time with your family and friends.
- **Make a calendar.** Write your schedule on a calendar. You can quickly see everything you need to do.

Why is it a good idea to plan a budget?

ACTIVITY LIFE SKILLS CRITICAL THINKING

Access Health Facts

Suppose you want to make a budget that will let you buy a new bicycle. Use health facts to make a sample budget.

1. **Identify when you might need health facts.** What bicycle will you buy? How much will it cost? How much can you save each week?
2. **Identify where you might find health facts.** List three sources you could use to find the information you need.
3. **Find the health facts you need.** Use the sources you listed. Collect the facts you need. Be sure you use reliable information.
4. **Evaluate the health facts.** Write a sample budget showing how you could save enough for the bicycle. Which source did you find most useful?

LESSON REVIEW

Review Concepts

1. **Describe** how government agencies and consumer advocacy groups help protect consumers.
2. **Explain** why budgets and schedules are important.

Critical Thinking

3. **Contrast** How is the Food and Drug Administration different from the Better Business Bureau?
4. LIFE SKILLS **Access Health Facts** What information could you get from the Consumer Product Safety Commission?
5. LIFE SKILLS **Manage Stress** A person might feel stressed because he or she has too many things to do. How can a schedule help a person do something about his or her stress?

CRITICAL THINKING

Use Communication Skills

Problem You just got a new bike helmet. The first time you wore it, the chin strap broke. What can you do?

Solution You can use communication skills to explain the problem to someone at the store. Using the steps on the next page can help.

Learn This Life Skill

Follow these steps to help you communicate when you have a problem with a product. The Foldables™ can help you follow the steps.

1 Choose the best way to communicate.

People talk, write, and use body language to communicate. Which is the best choice in this situation? You could speak to the manager at the store.

2 Send a clear message. Be polite.

Explain what happened. Tell the manager that the bicycle helmet broke the first time you wore it. Use I-messages to explain how this made you feel. Be sure your body language matches what you say.

3 Listen to each other.

The manager may ask you questions about what happened. Use the active listening skills you learned in Chapter 2.

4 Make sure you understand each other.

Answer any questions the manager asks you. Ask the manager if he or she understands what you want. How can you make sure you both understand each other?

Practice This Life Skill

With a partner, list some situations in which you might call a business about a product. Then act out how you could use communication skills in one situation. Be sure to use all four steps listed above.

LESSON 3

Planning a Healthful Community

You will learn . . .

- to identify the health needs in your community.
- how communities meet their health needs.
- ways that volunteers can contribute to the health of a community.

Vocabulary

- **disease prevention**, *E17*
- **disease control**, *E17*
- **sanitation**, *E19*
- **volunteerism**, *E20*

Did you ever think about what it takes to keep your community a safe and healthful place? Many people work to keep your community healthful. Ordinary citizens like you can also pitch in and lend a hand.

Community Health Needs

Your community has many health needs. The trash must be collected. The water must be safe to drink. The waste that goes down the drain must be treated. Stray animals must receive care. Buildings must be checked for safety.

Many of these needs are part of disease prevention. **Disease prevention** involves actions taken to prevent illness before it starts. Untreated water can carry pathogens. Treating the water keeps people from getting diseases from those pathogens.

Vaccines are another important part of disease prevention. If everyone in a community gets the vaccine for a disease, then the disease can't spread in that community.

Communities also need disease control. **Disease control** concerns actions taken to control the spread of illness. Suppose a student at school is ill. He or she visits the school nurse. The nurse sends the student home. This helps protect other students from getting ill.

Suppose an ill person works in a restaurant. He or she could spread pathogens to the food. Pathogens can also spread if a restaurant is not clean. Workers in communities need to be sure that restaurants and markets are clean so pathogens do not spread.

What is disease prevention?

ACTIVITY

Social Studies LINK

Mapping Disease

In 1854, a disease called cholera (KAH•ler•uh) struck London, England. A doctor named John Snow thought cholera was transmitted through water. He made a map showing the houses where people died from cholera. His map also showed the water pumps that people used. Look at the map below. Suppose a worker at the city health department asked you which pump the city should turn off to stop the disease. With a partner, write a dialogue between you and the health department. In the dialogue, suggest which pump to turn off. Explain your reasons. Perform your dialogue for the class.

Outbreaks
Pumps
Roads

▶ **Safe water and animal control are two needs many communities share.**

Meeting Community Health Needs

ACTIVITY

Art LINK

Design a Diorama

Design a diorama showing one way a health, sanitation, or recreation department helps a community's health. Use cutouts, clay, or other modeling materials to make your diorama. Share your diorama with your class.

Many people work to meet your community's health needs. Some work for the government. Others work for private groups. Suppose that a big storm struck your community. Your community's police and fire departments would respond. Hospitals would treat injured people. Other organizations would help people who lost their homes.

Health Department

Most communities have a health department. This department works on both disease prevention and disease control. The health department teaches people how to avoid illness. The department may offer low-cost or free vaccines, or *immunizations*. People in the health department also work to control diseases that are spread by insects.

Inspectors from the health department make sure restaurants and markets are clean and safe. Inspectors also make sure workers in restaurants do not have communicable diseases.

Pixtal/AGE Fotostock

▲ **Health department workers teach people how to stay healthy.**

Sanitation Department

Your community probably has a sanitation department. **Sanitation** is the protection of people's health by getting rid of trash and keeping places clean. Workers in your community's sanitation department clean the streets. Its workers pick up the trash. They may remove items that can be recycled. You will learn more about how communities deal with trash in Chapter 10.

Workers in the water and sewer department keep water clean. The department also deals with sewage. Sewage is usually sent to a sewage treatment plant where workers clean the sewage out of the water. The water is then released into a nearby river, lake, or ocean.

Recreation Department

Workers in your community's parks and recreation department care for parks, playgrounds, bike trails, and sports fields. This allows people in the community to have fun. It also gives people a chance to enjoy healthful activities.

Look around your home for recycled products. How many can you find? List the products you find. Make a bar graph showing how many products are made from recycled paper, plastic, metal, or some other material. You may find recycled materials in packaging, paper products, plastic bottles—even trash bags!

What does the water and sewer department do?

In many communities, the sanitation department runs recycling centers.

The Importance of Volunteers

ACTIVITY
BUILD Character

Clean Up a Park

Citizenship Choose a park or a playground in your neighborhood that needs to be cleaned. With your parents or guardian, organize a group of friends to clean up the park. Make sure that an adult supervises the cleanup and that everyone wears safety gear, such as rubber gloves. How else could you help your community?

Sometimes communities do not have enough money to pay someone to do all the jobs that need to be done. Volunteers can take over some of these important tasks. **Volunteerism** is the providing of a service without being paid.

Think about what needs to be done in your community. Is there a beach or vacant lot that needs to be cleaned? Is there a food pantry or a shelter that could use help collecting supplies?

Volunteers can also help people one-on-one. You could volunteer to help tutor a classmate in a subject. Do you have neighbors who are ill? They may welcome your help with chores.

Fuse/Getty Images

▲ You can help your community and have fun at the same time.

When you volunteer, you help your community and you help yourself as well. You can learn new things. You'll meet new people. Some of them could become good friends.

Volunteering can also improve your health. You'll feel better about yourself because you're helping other people and your community. Feeling good about yourself helps your mental and emotional health. Volunteering can be fun, too!

Why are volunteers important in a community?

▲ **What would happen if nobody picked up the trash?**

ACTIVITY LIFE SKILLS

CRITICAL THINKING

Analyze What Influences Your Health

Write a short story about what might happen if nobody took care of your community's health needs.

1. **Identify people and things that can influence your health.** Choose a group that helps keep the community healthful. For example, the sanitation department collects the trash.
2. **Evaluate how these people and things can influence your health.** How does the group affect your health? If the trash piled up, how would it affect your health?
3. **Choose healthful influences.** Be sure your story shows how important it is to meet the community's needs. Also include how the person or group you chose helps meet the needs.
4. **Protect yourself against harmful influences.** Your story should show how community health workers protect the community. Share your story with your classmates.

LESSON REVIEW

Review Concepts

1. **List** the health needs of a community.
2. **Identify** the different departments that help meet a community's health needs.
3. **Describe** how volunteers contribute to the health of a community.

Critical Thinking

4. **Analyze** How do immunizations protect the health of people in a community?
5. LIFE SKILLS **Analyze What Influences Your Health** How can volunteering influence your health?
6. LIFE SKILLS **Be A Health Advocate** How do volunteers communicate a healthful message?

Jill Braaten/McGraw-Hill Education

LESSON 4

Careers in Health and Fitness

You will learn . . .

- how to find out more about health careers.
- what jobs health workers perform.
- how to prepare for a health career.

Vocabulary

- **career**, *E23*
- **mentor**, *E23*

Do you think you might like a job related to health? When you find a job you enjoy and do it well, it's good for everyone, especially you.

Learning About Careers

When you work, you do something useful for the community and yourself. You earn money that you need. If you do your job well, you feel good about yourself. You might even have fun!

A **career** is the work that a person prepares for and does throughout life. There are many careers related to health. You can find information on health careers at the library. You can also talk to a mentor who has a career you're interested in. A **mentor** is a responsible person who helps a younger person. In some cases you can actually go to work with the mentor and watch as the person does his or her job. This allows you to see if you would like that health career.

How can a mentor help you?

Write About It!

Write About a Career With your teacher, arrange for people who have careers in health care to talk to your class. Write down questions that you want to ask each person. After they speak, evaluate each career. What did you learn about the job? Has your interest in the job increased or decreased? Why? Share your thoughts with the class.

Echo/Getty Images

◀ Spend time with someone at work to find out about his or her job.

Careers in Health Care

In 2000, there were 598,000 physicians in the United States. There were more than 2 million registered nurses!

You can find people with health careers throughout your community. Some health workers work in hospitals and doctors' offices. Other health workers take care of people in emergencies. There are health workers in your school and your community, too.

Hospitals and Doctors' Offices

Hospitals and doctors' offices employ many health workers. Nurses and doctors, also called *physicians*, care for patients. Dieticians manage hospital kitchens. They make sure patients get the right foods. Pharmacists get medicines ready for patients.

Physical therapists help people use body parts that have been harmed by illness or injury. Physical therapists often teach people how to do exercises. These exercises help people recover from their injuries.

Emergencies

Some health workers learn how to help in emergencies. *Emergency medical technicians* (EMTs) give first aid to people when an emergency occurs. They take care of people on the way to the hospital.

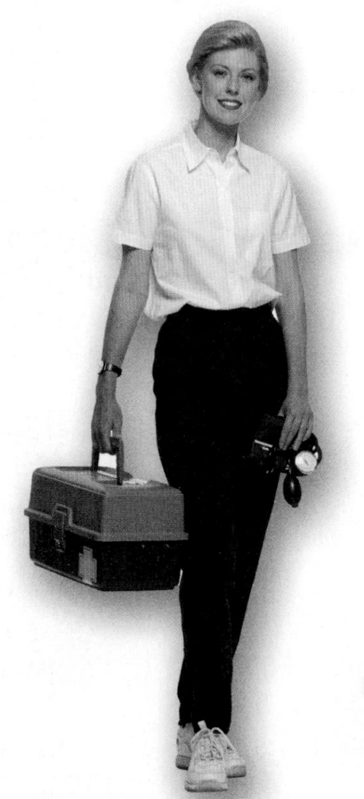

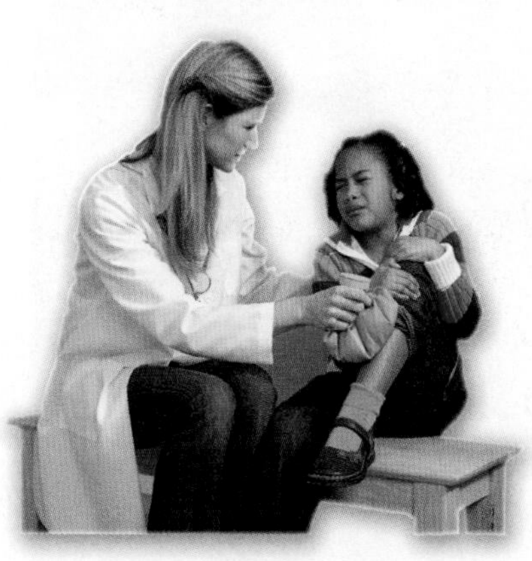

(l)Comstock/SuperStock, (c)Eclipse Studios/McGraw-Hill Education, (r)Fuse/Getty Images

▲ **What do these people's jobs have in common? How are they different?**

School

The *school nurse* offers care for students and teachers who become ill or injured at school. Many school nurses also teach about health. Physical education teachers help students learn fitness skills. The *school psychologist* helps students solve personal problems. Your school may also have counselors that help students avoid drugs and alcohol. Other counselors can help you learn about careers.

Community

Some nurses and physical therapists work with people in their homes. Personal trainers help people exercise safely. Exercise teachers and nutritionists teach classes. These classes teach people how to exercise and eat healthfully. People with these careers may work at fitness centers. Governments and private groups also employ health workers. These people help others resolve health-care problems.

What kinds of health careers are available at a hospital?

▲ **Tests can help you discover your career skills and interests.**

Design Pics/Ron Nickel/Getty Images

ACTIVITY

On Your Own

FOR SCHOOL OR HOME

Design a Dream Job

Write and illustrate a short book or presentation about a health career you are interested in. Choose the health-related job you would most like to have. Where would you work? What would you do? Why would you want the job? Include your answers in your book or presentation.

Preparing for a Career

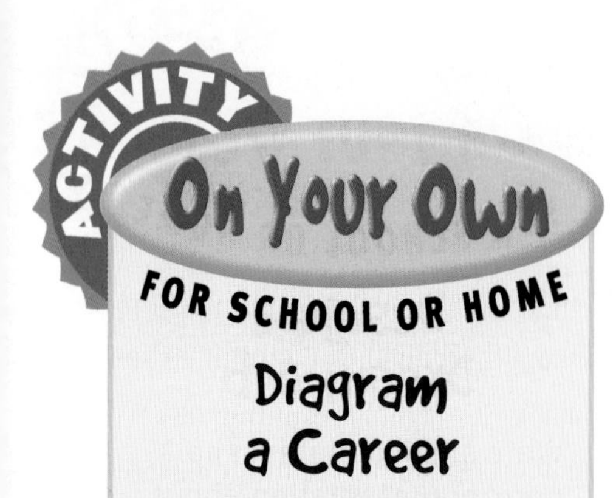

Diagram a Career

Choose one career from the chart. Draw a diagram showing all the steps a person would need to follow to prepare for the career. Find out more about the training required for the career. Be sure your diagram includes all the different kinds of school and training needed for the career. Combine your diagram with your classmates' diagrams in a class career book.

A few health-care careers are listed below. Which ones interest you?

HEALTH CAREERS

Career	What They Do	Typical Education/Training
Dental Hygienist	Clean and examine teeth and gums, take and develop X rays.	Degree from school of dental hygiene and license. School and training last 2–4 years.
Dentist	Treat problems of the teeth and gums.	Degree from dental school and license. School and training last about eight years.
Dietician	Plan food and nutrition programs.	College degree. School and training last at least four years.
Home Health Aide	Help people in their homes. Often give people medicine and and help them move.	Training program and test. Many people can become health aides right after high school.
Nurse Practitioner	Treat some conditions and prescribe drugs without the supervision of a doctor.	Advanced nursing courses and license. School and training last 5–6 years.
Personal Trainer	Teach people how to exercise safely. Help people meet physical fitness goals.	Certification in the type of training they do; some need a college degree. A degree takes about four years.
Pharmacist	Provide drugs prescribed by physicians and provide information about drugs.	Degree from school of pharmacy and license. School and training last at least six years.
Physician	Diagnose and treat people suffering from illness or injury.	Degree from medical school, additional training, and license. School and training last 11–16 years.
Psychologist	Provide mental health care.	Graduate school degree and license. School and training last 9–11 years.

There are many other health careers as well. Some people make sure that doctors' records are accurate. Others study public health issues or work for community health departments. Others do research to find new medicines.

If you think you'd like a career in health care, consider everything about the job. Where would you work? What would you do each day? What kind of training would you need?

For many health careers, you need to study science. It's also important to study writing and communication skills. In any job, you will need to communicate with other people. Math is also important in many health careers.

Also think about your interests. Do you like working with people? Would you rather spend your time in a doctor's office or in a laboratory?

People with health careers are sometimes called health helpers. All of them help keep people and communities healthy. How can you be a health helper?

What does a nurse practitioner do?

ACTIVITY

LIFE SKILLS

CRITICAL THINKING

Set Health Goals

1. **Write the health goal you want to set.** Think of five health careers that you would like to learn about. Write them into a Health Behavior Contract.
2. **Explain how your goal might affect your health.** How can learning about careers affect your health? Record your ideas.
3. **Describe a plan you will follow to reach your goal. Keep track of your progress.** Each time you learn about a career, put a check mark next to it. Try to learn about one career each day for a week.
4. **Evaluate how your plan worked.** At the end of the week, look at your chart. Did you learn about the health careers you selected? If not, explain what you can do to succeed next time.

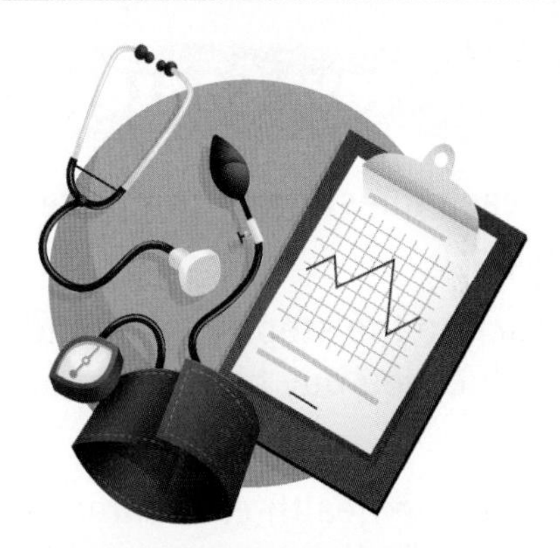

LESSON REVIEW

Review Concepts

1. **Identify** ways to learn about health careers.
2. **List** one health job that can be found in each of the following: a hospital or doctor's office, an emergency situation, a school, and in the community.
3. **Describe** the training required for three health-care careers.

Critical Thinking

4. **Evaluate** What job characteristics would you consider when choosing a health career?
5. LIFE SKILLS **Set Health Goals** How can spending the day with a mentor help you identify health career goals?

CHAPTER 9 REVIEW

Use Vocabulary

appeal, *E6*
budget, *E12*
career, *E23*
consumer, *E5*
disease prevention, *E17*
media, *E6*
volunteerism, *E20*

Choose the correct term from the list to complete each sentence.

1. A plan for spending and saving money is a __?__.
2. The providing of a service without being paid is __?__.
3. Stopping illness before it starts is __?__.
4. Someone who judges information when buying products and using services is a __?__.
5. Sources of news and information are the __?__.
6. A statement used to make consumers want to use a product or service is a(n) __?__.
7. The work that a person prepares for and does throughout life is a __?__.

Review Concepts

Answer each question in complete sentences.

8. Describe the steps you should take if a product you buy breaks or does not work.
9. Name five health-care careers. Explain what each does.
10. List two government agencies that help consumers. What does each do?
11. What should you include when making a budget?
12. What types of media might feature advertisements and appeals?

Reading Comprehension

Answer each question in complete sentences.

Your community probably has a sanitation department. Sanitation is the protection of people's health by getting rid of trash and keeping places clean. Workers in your community's sanitation department clean the streets. Its workers pick up the trash. They may remove items that can be recycled.

13. What is sanitation?
14. What does the sanitation department do?
15. How does the sanitation department help a community's health?

Critical Thinking/Problem Solving

Answer each question in complete sentences.

Analyze Concepts

16. Suppose you want to buy a new pair of shoes. How might the media influence your choice?

17. A community needs a new playground. There isn't enough money to pay someone to build it. How might volunteers in the community help?

18. Give an example of a reliable source of health information. How do you know that the information is reliable?

19. Why are consumer advocacy groups usually more effective than an individual?

20. Which health career most interests you? Explain why.

Practice Life Skills

21. **Use Communication Skills** You bought a CD, but when you play it, you find that it is scratched and skips songs. How could you communicate the problem to someone who can help you?

22. **Make Responsible Decisions** You see an ad for a new sports drink. It looks good. The ad claims it will make you play sports better. You aren't sure the claim is true. What should you do? Use the *Guidelines for Making Responsible Decisions*™ to help you decide.

Read Graphics

Look at the budget that Emma made. Use it to answer questions 23–25.

Income	Expenses
\$4.00 Allowance \$2.00 Chores (helped shovel snow off the driveway) \$10.00 Birthday money from Grandma	\$5.50 Go to movies \$2.00 Buy pencils for school \$4.00 Save to buy a new pair of jeans \$1.50 Buy a birthday card for Uncle Paul \$2.00 Buy snacks after school
Total: \$16.00	Total: \$15.00

23. Emma has enough money to cover her expenses this week. If she did not have birthday money from Grandma, would she still have enough money?

24. Emma's class is going on a field trip to the museum. Each child needs \$5.00. How could Emma change her budget so she has enough money to go?

25. What could Emma do to improve her budget?

CHAPTER 10

Environmental Health

Lesson 1 • Your Environment. *E32*

Lesson 2 • Protecting Water and Land *E40*

Lesson 3 • Conservation. *E46*

Lesson 4 • A Positive Environment. *E50*

What Do You Know?

What can you do to protect the environment? Read each statement. On a separate sheet of paper, write **yes** if the statement describes a way to protect the environment. Write **no** if it does not.

__?__ **1.** Put litter in a trash can.
__?__ **2.** Pour old paint and cleaning supplies down the drain.
__?__ **3.** Leave lights on all the time.
__?__ **4.** Keep the volume down when you listen to music.
__?__ **5.** Walk or ride a bike instead of riding in a car when you can.
__?__ **6.** Buy products with extra packaging.

Statements 1, 4, and 5 are ways to protect the environment. What other ways can you think of? Read **Environmental Health** to learn more about why it is important to protect the environment and how you can help.

LESSON 1

Your Environment

You will learn . . .

- factors that lead to a healthful environment.
- how air and noise pollution affect the health of the environment.
- how groups and individuals can reduce air and noise pollution.

Vocabulary

- **pollution**, *E33*
- **pollutants**, *E33*
- **air pollution**, *E34*
- **noise**, *E35*

The air, water, land, and even the sounds around you are all part of your environment. Activities people do every day, such as driving cars, can harm the environment. But there are steps people can take to help protect the environment.

Our Natural Resources

The environment is everything that is around you. It includes people, plants and animals, lakes and oceans, the air you breathe, and the soil under your feet. You depend on the environment for the water you drink and the food you eat. These things from nature that people need or use are called *natural resources.*

All parts of the environment depend on each other. Something that affects one part of the environment can affect other parts, too. An unhealthful environment can harm the people and animals that live in it.

Suppose a lake becomes polluted. **Pollution** is when land, water, or air contains harmful substances, making them unhealthful to the environment and people. These substances are called **pollutants**. Pollutants can get into the fish that live in the lake. Then birds and animals eat the fish. The pollutants can harm these animals as well. They can also harm plants that grow near the lake. Pollution can harm people who drink the water from the lake or swim in it.

What is pollution?

Write About It!

Write a Descriptive Essay
Write a short essay describing your environment. For a day, pay close attention to your environment. Write about the air, the land, the water, and everything else around you. Describe what is healthful about your environment. Also describe any pollution you notice.

▼ **Everything around you is part of your environment.**

Air Pollution

ACTIVITY

Science LINK

Test for Pollutants

Spread a little petroleum jelly on two index cards. Leave each card in a different place. You might put one in your classroom and one outside. Choose places where the card won't get wet. After a few days, check each card. Use a hand lens if you can. What pollutants did you find? Which card was dirtier? Why do you think it was dirtier?

The air you breathe can become polluted. **Air pollution** occurs when harmful particles and gases get into the air. Most air pollution comes from fuels. Cars, trucks, factories, and power plants burn these fuels to get power. People also burn these fuels to heat houses. Burning these fuels puts pollutants into the air. Three major air pollutants are carbon monoxide, sulfur dioxide, and smog.

Carbon monoxide (muh•NOK•sighd) is an odorless gas produced by cars, trucks, and buses. Breathing it keeps your body from getting the oxygen it needs. At high levels it can cause death.

Sulfur dioxide (SUL•fuhr digh•OK•sighd) comes from power plants that burn coal and oil. It can cause breathing problems. It also mixes with water vapor in the air to form acid rain. *Acid rain* is rain or snow that contains harmful chemicals. Acid rain can damage forests and harm living things in lakes and streams.

Smog is a haze created by sunlight shining on polluted air. Smog contains a gas called ozone (OH•zohn). Ozone can cause coughing and harm a person's lungs.

Indoor Air Pollution

Air pollution occurs indoors, too. Secondhand smoke and radon are two major causes of indoor air pollution.

You learned in Chapter 7 that secondhand smoke can increase your risk of disease. Your risk can be reduced if your family's guidelines don't allow smoking in your home.

▶ **Sunlight striking polluted air can create smog.**

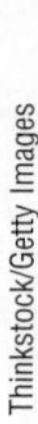

Radon (RAY•dahn) is a gas created below the surface of the ground. It can seep into a house through floors and walls in the basement. Over time, exposure to high levels of radon can lead to cancer. Your home can be tested for radon. Sealing cracks in the walls and floor of a basement can reduce the level of radon in a house.

Noise Pollution

Noise is loud or constant sound. The roar of traffic and the scream of a police siren are noise. Even your favorite music can be noise if someone plays it loudly when you are trying to sleep.

Loud noise can harm cells in the ear. If you hear loud noise over long periods of time, you could develop hearing loss.

Noise can also cause stress. It can make you feel nervous and make your heart beat faster. It can be hard to concentrate when it's noisy.

Many communities try to protect people from noise. Some have laws that ban loud music after a certain time at night. Many communities put barriers along highways to reduce noise from traffic. Some buildings have soundproof walls.

You can help protect yourself and others from noise pollution. Make sure that you set the volume on radios, TVs, and CD players at a low level. Keep the sound low even when you listen to music with headphones. Wear earplugs when you go to loud concerts, too.

What is noise?

ACTIVITY

LIFE SKILLS

CRITICAL THINKING

Use Communication Skills

Suppose that there is an air or noise pollution problem in your community. Make a plan to tell someone about the problem.

1. **Choose the best way to communicate.** Identify someone who can help you. What is the best way to communicate with that person? Should you make a phone call or visit the person? Write the name of the person and how you will contact him or her.
2. **Send a clear message. Be polite.** Decide what you will say. Be sure to use I-messages. Write an outline of what you want to communicate.
3. **Listen to each other.** What listening skills would you use? List some questions the person might ask you.
4. **Make sure you understand each other.** Write an answer to each question. Make sure you explain yourself clearly.

▲ **What could cause noise pollution in your community?**

Cleaning Up the Air

ACTIVITY

Social Studies LINK

Write a Law

With a partner, write a law about reducing air pollution. Suppose you are part of your state government. How would you reduce air pollution? Decide what the law will say. How will it help the air? Write your law and design a brochure to persuade people to vote for your law.

Many organizations work to control pollution and protect the environment. Government agencies such as the Environmental Protection Agency (EPA) enforce laws to reduce pollution. Workers in health organizations such as the American Lung Association and the American Cancer Society help teach people about the effects of pollution. Environmental groups teach people about the environment. They also work to protect it. Individuals can also help to prevent pollution and protect the environment.

Clean Air Laws

The U.S. government passed the Clean Air Act in 1970. The *Clean Air Act* is a law that limits the amount of pollution that can be in the air. Because of this and other laws, the air has become cleaner.

Machines around the country sample the air each day. The machines tell officials how much of certain kinds of pollution are in the air. For example, if the level of carbon monoxide is too high, the machine will record it. Then someone must find a way to lower it.

Your state has its own laws about the environment. Your community might have a community environmental plan, too. The plan determines the way the community will reduce pollution, get rid of trash, and protect parks and forests.

Laws such as the Clean Air Act have made factories and power plants reduce the pollutants they release into the air. Cars, trucks, and buses are now made with filters. The filters get rid of some pollutants before they get into the air. In many states, cars must be inspected every year to make sure these filters are working.

▲ **Motor vehicles are responsible for nearly half of the toxic air pollutant emissions in the United States. Source: EPA**

The fuels used in cars have changed, too. Gasoline used to contain lead. The lead was added to make engines run better. But this polluted the air with lead. Lead is a poisonous metal that can harm people's health. Laws were passed to remove lead from gasoline. The amount of lead in the air has dropped during the past twenty years.

▲ **Planting trees helps improve your environment.**

What You Can Do

Individuals can help keep the air clean, too. Here are some ways you can reduce air pollution.

- **Use electricity sparingly.** Most power plants burn fuel to make electricity. The less electricity you use, the less power they need to produce.
- **Walk or ride your bike when possible.** Fewer car trips mean less air pollution.
- **Do not smoke.** Smoke causes indoor air pollution.
- **Plant trees.** Trees produce oxygen from carbon dioxide found in the air.
- **Do not burn trash.** Burning trash creates smoke that can make people's eyes and throats sting.

List two reasons that motor vehicles produce less air pollution now than they did years ago.

LESSON REVIEW

Review Concepts

1. **Describe** an environment that is healthful.
2. **Explain** how air and noise pollution affect your environment.
3. **List** ways that the government, companies, and individuals can help reduce air pollution.

Critical Thinking

4. **Synthesize** Explain how electricity use and air pollution are related.
5. LIFE SKILLS **Use Communication Skills** You want to report a problem to your state environmental agency. How could you identify the best person to contact?

KidStock/Blend Images/Getty Images

Learning LIFE SKILLS

CRITICAL THINKING

Set Health Goals

Problem Your sister complains that you play your music too loudly. She gets angry when the music keeps her from concentrating on her homework. You know that loud music could harm your hearing. How do you respond?

Health Behavior Contract

Name ______________________ **Date** ______________

Health Goal: I will play my music at a safe level.

Effect on My Health: Keeping music at a safe level will help protect my hearing. It also shows respect for my sister and other people in my family.

My Plan: When I turn on my CD player in the evening, I will step outside the door of my room to check whether it's too loud. I will ask my sister if it is soft enough not to bother her. I will put a star next to the days of the week that I stick to my plan.

My Calendar

Sun.	Mon.	Tues.	Wed.	Thurs.	Fri.	Sat.

How My Plan Worked: I will talk about how my plan worked. For example, if I do not keep my music at a safe level one day, I will tell why.

Solution You can set a goal to keep your music at a safe level. Follow these steps for setting a health goal.

Learn This Life Skill

Follow these steps to help you keep noise at a safe level. The Foldables™ can help.

1 Write the health goal you want to set.

Your goal is, "I will keep noise at a safe level." Write this goal into your Health Behavior Contract.

2 Explain how your goal might affect your health.

How will reducing the amount of noise affect your health and the health of others around you? Write your answer in your contract.

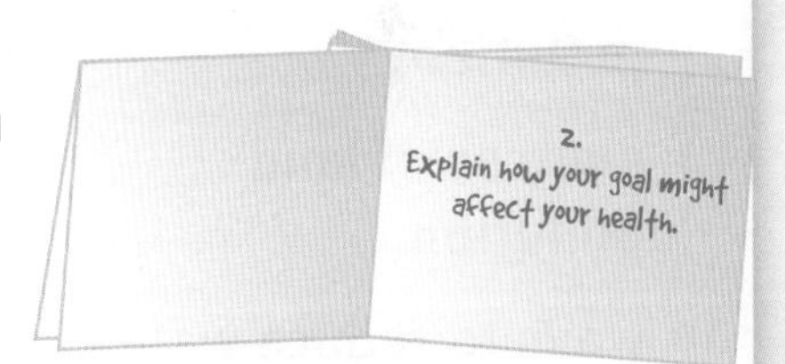

3 Describe a plan you will follow to reach your goal. Keep track of your progress.

Make a Health Behavior Contract. Use the one shown on the previous page as a guide. Describe how you will communicate your plan to your family. Check as you go along to see whether you're making progress.

4 Evaluate how your plan worked.

Describe how the plan worked. Did you succeed in reducing the amount of noise you make? If your plan did not succeed, evaluate your plan. How might you improve it?

Practice This Life Skill

Think of an action you can take to help reduce pollution. For example, you could be responsible for recycling waste material from your home. Use the steps for setting health goals to help you make a plan. Write a Health Behavior Contract for your plan.

LESSON 2

Protecting Water and Land

You will learn . . .

- sources of water pollution.
- sources of land pollution.
- ways to reduce water and land pollution.

Vocabulary

- water pollution, *E41*
- land pollution, *E42*

Water and land are two of the most important natural resources. We need to keep them free of pollution. Thanks to the government, private groups, and individuals, our land and water are getting cleaner.

Water Pollution

Design a Safety Poster

Design a poster to show how to use chemical products safely. In a grocery store or home center, look at products used on plants or lawns. Choose one product. What safety instructions are on the label? What could you do to use the product in ways that do not pollute water or land?

The water in your home might come from a lake or river. It might come from a well. Wherever it comes from, it needs to be clean. **Water pollution** is when water contains harmful substances, making it unhealthful to the environment and people. Pollution can make water not safe to drink.

Rain water often flows off the land into streams, rivers, and lakes. This is called runoff. Chemicals that people use to kill weeds and insects sometimes get into the runoff. As these chemicals build up, they can harm people and animals.

Chemicals can also wash into water from factories. Sometimes factories store wastes in containers. If the containers leak, the waste can run off into the water.

Sewage is waste that is flushed down the toilet or poured down the drain. The waste may contain bacteria or viruses. Sewage can also contain harmful chemicals. These can include paint and cleaning products you might find at home.

What is sewage?

Red Tide

Algae are microscopic single-celled plants and can grow quickly. *Red tides* are harmful algae in the ocean turning it a red color. Toxins from red tides pollute ocean water and air. Strong winds can carry red tide toxins further inland affecting air quality.

Signs of red tide include dead fish and birds on the beach and a foul smell in the air. Eating shellfish from a red tide can make you ill. Breathing air with red tide toxins can cause coughing and burning in the eyes, nose, and throat.

Red tides occur near the ocean in Texas, Florida, and the North Atlantic. Check for red tide warnings. If you feel ill due to a red tide get into an air-conditioned car or building. Get prompt medical attention if you have problems breathing or have eaten contaminated shellfish. Be extra careful if you have asthma.

▼ **Paints and cleaners can contain chemicals that pollute water. Don't pour them down the drain.**

Bob Coyle/McGraw-Hill Education

Land Pollution

Health Online

Watch the Waste

Many communities must decide how to clean up polluted land. Research ways communities cleaned polluted land. Write a report on what you learn using the e-Journal writing tool.

Like air and water, soil is part of the environment. Most plants can't grow without it. Farmers use it to grow food. **Land pollution** is when land contains harmful substances that make it unhealthful to the environment and people. One kind of land pollution comes from wastes left on the ground. Another comes from wastes that leak into the ground.

Litter is trash that people drop onto the ground. You probably see litter often. Candy wrappers, soda cans, and food can all be litter. Litter makes an area look dirty and ugly. It can also attract rodents and insects. These animals carry pathogens. Litter can also wash into rivers and oceans when it rains.

Open dumps are areas where waste is dumped and left on the ground. Open dumps look and smell bad. They attract pests. The waste they contain can seep into the soil. Chemicals and pathogens get into the soil. Then they spread to plants, animals, and buildings.

▼ **When people dig mines, they can leave behind piles of rock and chemicals.**

Robert Kamilli/USGS

Hazardous waste is waste that contains substances that are harmful to people or animals. Batteries and used motor oil are examples of hazardous waste. Some factories produce hazardous waste. The waste can cause land pollution if it is not stored or disposed of safely. Mining can also produce hazardous waste. Minerals are separated from other rock. The leftover rock, called slag or tailings (TAY•lings), can contain pollutants. This rock piles up into hills. Hazardous waste in these hills can be spread by wind and rain.

What is land pollution?

Write About It!

Write a Narrative Story Write a short story about how to stop litter. Be creative in your story. Make your story funny so it will catch other people's attention.

CRITICAL THINKING

Make Responsible Decisions

Suppose you and your family are having a picnic in the park. Suddenly, it starts to rain. Should you leave your trash behind and run for cover? You know that the parks department will clean up the mess sooner or later. What is the responsible decision? With a partner, role-play the situation.

1. **Identify your choices.** You could stay dry and leave your trash behind. What else could you do? Role-play the discussion with your partner.
2. **Evaluate each choice. Use the *Guidelines for Making Responsible Decisions™*.** Review each choice using the Guidelines. One person should ask the other person the questions.
3. **Identify the responsible decision. Check this out with your parent or trusted adult.** What is the responsible action when it begins to rain on your picnic? You and your partner should role-play how you could agree on the decision.
4. **Evaluate your decision.** Explain why you made the decision you did. Each person should do part of the explaining.

Guidelines for Making Responsible Decisions™

- **Is it healthful?**
- **Is it safe?**
- **Does it follow rules and laws?**
- **Does it show respect for myself and others?**
- **Does it follow family guidelines?**
- **Does it show good character?**

Reducing Water and Land Pollution

CAREERS

Recycling Plant Worker

People who work at recycling plants do many different jobs. Some collect and sort recycled materials. Other workers prepare the recycled materials to be used in new products. Other workers are engineers who decide how the recycled materials can be used. These workers help keep a community's environment clean.

Years ago, people in many communities dumped untreated sewage into lakes, rivers, and oceans. The sewage contained human waste and harmful chemicals. People in many communities used open dumps. Today, there are laws to prevent and control pollution on land and in water.

The Clean Water Act limits the amount of pollutants that can be released into water. The sewage of most communities must pass through sewage treatment plants. These plants clean many harmful substances out of sewage. This makes it safer to release the sewage into rivers, lakes, or oceans.

Some chemicals are safe in small amounts. They can be harmful in larger amounts. The Safe Drinking Water Act limits how much of these chemicals can be in drinking water. Drinking water must be tested. The tests make sure that the water doesn't have too much of any of these chemicals.

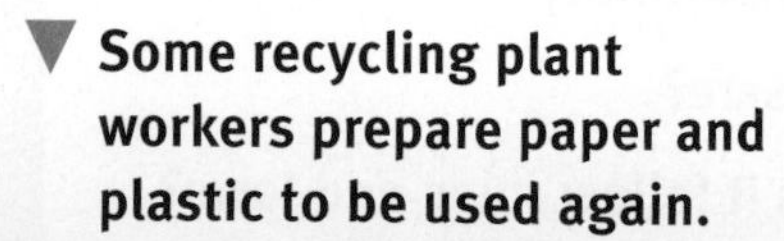

▼ Some recycling plant workers prepare paper and plastic to be used again.

Reduce Water Pollution

You can do your part to help keep water clean. Here are some rules to follow.

- **Dispose of chemicals properly.** Don't pour paint, paint thinner, or strong cleaners into the toilet or down the drain. Take them to a hazardous waste collection center. Your community may have a special hazardous waste collection day.
- **Dispose of trash properly.** Don't dump trash into streams, lakes, rivers, or the ocean.
- **Volunteer to help clean up** along rivers, lakes, or other bodies of water in your community.

Reducing Land Pollution

Today cleaner methods are often used to dispose of wastes. A *landfill* is a pit where trash is buried and covered with soil and clay. Wastes can still leak into the ground, though. *Sanitary landfills* have liners of plastic or clay under them. The liners help prevent wastes from leaking.

A community may send its waste to an incinerator (in•SIN•uh•ray•tuhr) instead of to a landfill. An *incinerator* is a furnace that burns trash and other waste. Filters on incinerators keep harmful substances out of the air.

What You Can Do

Here are some ways you can help reduce land pollution.

- **Don't throw litter** on the ground. Throw it into a garbage can or litter basket.
- **Take part** in a recycling program. Recycle paper, metal cans, and glass and plastic containers.

List two ways that communities get rid of waste.

ACTIVITY

On Your Own

FOR SCHOOL OR HOME

Dispose of Hazardous Wastes

Find out what system your community has for getting rid of hazardous waste. This waste includes paints, cleaners, motor oil, and batteries. Make a chart that shows these materials and the proper way to dispose of them. Hang the chart where your family can see it.

LESSON REVIEW

Review Concepts

1. **List** three causes of water pollution.
2. **Identify** three causes of land pollution.
3. **Explain** how communities work to reduce water and land pollution.

Critical Thinking

4. **Evaluate** How have the Clean Water Act and the Safe Drinking Water Act affected people's health?
5. LIFE SKILLS **Make Responsible Decisions** You have some oil-based paint to throw out. The easiest solution is to pour it down the drain. Is this a responsible decision? Explain your answer.

LESSON 3

Conservation

You will learn . . .

- ways to help conserve water.
- ways to help conserve energy.

Vocabulary

- conservation, *E47*
- nonrenewable resource, *E48*
- fossil fuel, *E48*
- renewable resource, *E48*

The energy in wind, sunlight, and falling water can be used to make electricity. Using them helps save other natural resources. It's important to save resources so they will last for a long time.

Conserving Water Resources

Conservation is the protection and careful use of forests, rivers, minerals, and other natural resources. Many natural resources are limited. Fresh water is one limited resource. There is only so much fresh water on Earth. When you conserve water, you help make sure that there will be enough for everyone to use.

You use water every day. You drink it, cook with it, and take baths and showers in it. Your family may use it to water the lawn and wash the kitchen floor. These tips can help you conserve water.

Math LINK

Calculate Water Usage

Most people use between 80 and 100 gallons of water each day. Suppose there are 85 people living in an apartment building. About how much water is used by the people in the building each day?

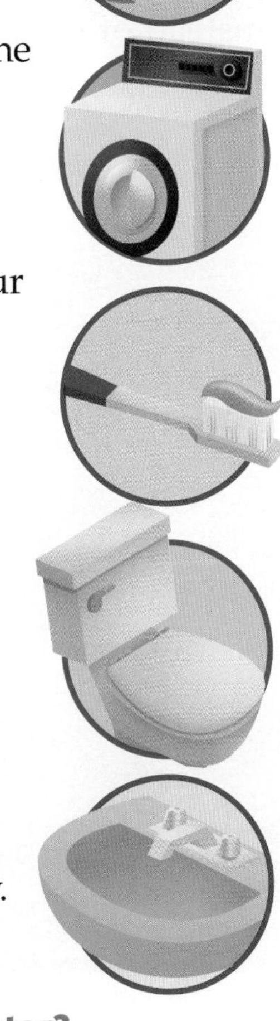

- **Take shorter showers** instead of baths. You'll save as much as 20 gallons of water.
- **Run only full loads** in the washing machine or dishwasher. These machines use anywhere from 16 to 35 gallons of water each time you run them.
- **Don't run the water** while you brush your teeth, wash the dishes, or put soap on your face. You can save more than five gallons of water just by turning off the water while you brush your teeth!
- **Don't use the toilet** as a garbage can. Throw away tissues or other items in the trash instead of flushing them down the toilet.
- **Tell your parents or guardian** if you see a leaky faucet or pipe. A leaky faucet can waste 100 gallons of water each day.

What does it mean to conserve water?

Conserving Energy

Energy is the ability to do work. It also refers to the fuels we use for heat and electricity. People need energy to run cars and trucks, too.

Most of the energy sources we use today are nonrenewable. A **nonrenewable resource** is one that cannot be replaced within a reasonable amount of time. The fossil fuels that we use are nonrenewable resources. **Fossil fuels** are coal, oil, and natural gas burned to make energy.

A **renewable resource** is one that can be replaced within a normal lifetime. Renewable energy resources include energy from the sun *(solar power)*, from the wind *(wind power)*, and from falling water *(hydropower)*. People are working to find ways to use these resources instead of fossil fuels.

There are two main reasons to use renewable resources instead of nonrenewable resources. First, renewable resources may not run out. We may never use up our supply of sunlight, wind, and falling water. Second, solar power, wind power, and hydropower are "clean" energy sources. They do not produce pollution.

Today, most of our energy still comes from nonrenewable fossil fuels. Conserving energy will help make those fuels last longer. It will also help reduce pollution.

ACTIVITY

Art LINK

Make a Conservation Poster

Use poster board and markers to draw a poster that encourages people to conserve energy. Include a slogan, tips, cartoons, photographs—anything that helps you get your message across. Share your poster with your classmates.

▲ **People have used water for power for many years.**

Conservation Tips

Here are some ways you can conserve energy.

- **Let daylight shine** into the rooms of your home. Use natural light instead of turning on a lamp.
- **Turn off lights** when you're not in the room.
- **Encourage your family** to use compact fluorescent (flaw•RES•uhnt) bulbs. These bulbs use less energy than regular light bulbs.
- **When it's cold, put on an extra sweater** instead of turning up the heat.
- **When it's hot, use a fan** instead of air conditioning.
- **Dry your clothes outside** on a clothesline instead of using a clothes dryer.

Practice Healthful Behaviors

Make it a habit to conserve energy and resources. Make a chart to help you.

1. **Learn about a healthful behavior.** With a partner, brainstorm a list of ways you can conserve energy in the home. Choose one to focus on.
2. **Practice the behavior.** Each partner should make a chart describing the behavior you chose. Post it in your home for everyone to see. Every time someone uses the behavior you chose, put a check mark on the chart.
3. **Ask for help if you need it.** Ask your parents or guardian to remind you to follow the behavior.
4. **Make the behavior a habit.** Keep marking your chart when anyone uses the behavior. How well is your family conserving energy?

LESSON REVIEW

Review Concepts

1. **Identify** five ways to conserve water.
2. **List** five ways to conserve energy.

Critical Thinking

3. **Compare and Contrast** How are renewable and nonrenewable resources the same? How are they different?
4. LIFE SKILLS **Practice Healthful Behaviors** Who could help you conserve energy around your home? How could this person help you?
5. LIFE SKILLS **Be a Health Advocate** Your sister leaves the water running while she brushes her teeth. How could you encourage her to conserve water?

LESSON 4

A Positive Environment

You will learn . . .

- what it means to precycle, reuse, and recycle.
- how to help make your environment positive.

Vocabulary

- **precycle**, *E51*
- **reuse**, *E51*
- **recycle**, *E51*
- **positive environment**, *E52*
- **visual environment**, *E53*

When you use something again instead of throwing it away, you improve your environment. You create less trash. This is just one way you can make your environment a healthful, positive place to live.

Precycle, Reuse, and Recycle

Each year, Americans throw away about 300 billion pounds of trash. Every time you throw an item away and buy a new one, you use resources. Most of these resources are nonrenewable. You can help make resources last longer.

Precycle

The first step in conserving resources is to precycle. To **precycle** is to take actions to reduce waste. To do this, choose items with less packaging. For example, instead of buying applesauce in six separate plastic cups, you can buy one large glass jar. Bring a cloth bag to the store to put your groceries in instead of using the store's plastic or paper bags.

Reuse

When you **reuse** something, you use it again instead of throwing it away and buying a new one. You could rinse out peanut butter or mayonnaise jars to use for storing items. When you receive gifts, save the boxes and wrapping paper to use again. Think of all the items you throw away in a day. How many of them could you use again?

Recycle

If you can't reuse something, try to recycle it instead of throwing it away. To **recycle** means to change waste products so that they can be used again. For example, you can put a newspaper into a recycling bin instead of throwing it away. It then goes to a paper recycling plant. The plant makes it into new paper. That paper might be the paper you use in your computer printer.

What does it mean to precycle?

MAKE a Difference

Reducing Trash

Students from Brown Elementary School in Denver, Colorado, discovered that half the material found in a landfill was recyclable. They wanted to reduce that amount. They added plastic and aluminum to the school's recycling program. They traded books and toys with each other instead of throwing them away. They started a worm box to make compost for gardens from their trash. What could you do to reduce trash in your school?

A Positive Environment

ACTIVITY
BUILD
Character
Volunteer to Help

Citizenship Is there something you can do to benefit your community's environment? Does a vacant lot or park need cleaning up? Could playground equipment use a new coat of paint? Does an ill neighbor need someone to run an errand? With a parent or guardian's permission, volunteer your time. You'll help others and feel good about yourself.

Do you live in a positive environment? A **positive environment** is an environment that promotes physical, mental and emotional, and family and social health. In a positive environment, you feel safe. You feel good about yourself. You know that there are people around you who support you and care about your feelings. Your friends, family, and other important people in your life encourage you to do your best. They are happy when you succeed. They do not make fun of you or put you down.

What You Can Do

How can you help make the environment positive for others? Here are some ideas.

- **Compliment other people when they do something positive.** Congratulate a friend who made the softball team. Praise a classmate who got an A on a report or test. Tell your parents or guardian that you enjoyed the meal they cooked. Tell your brother or sister that you like his or her new outfit. Compliments encourage people to do well.

Keeping parks and playgrounds clean and safe helps make a community's environment positive.

- **Avoid saying negative things about other people.** Put-downs won't make you a better person, and they make other people feel bad.
- **Make your home environment positive.** If you see something that needs to be done, do it without being asked. Make sure that you finish your chores. Help other people with their chores. Clean up after yourself.
- **Help make the visual environment positive.** The **visual environment** is everything a person sees. Help keep your yard or the front of your building neat. Clean up trash in your neighborhood. Don't put graffiti on walls. When your visual environment is attractive, you feel better about where you live.

What is a positive environment?

Be a Health Advocate

With a group, perform a skit to teach others about precycling, reusing, and recycling.

1. **Choose a healthful action to communicate.** You've learned that precycling, reusing, and recycling are healthful actions. Decide which actions you will show in your skit.
2. **Collect information about the action.** Team up with classmates. Gather information about the health benefits of choosing to precycle, reuse, and recycle.
3. **Decide how to communicate this information.** Write a skit about the benefits of precycling, reusing, and recycling.
4. **Communicate your message to others.** Perform your skit for your classmates. Show them how they can help the environment, their community, and themselves.

LESSON REVIEW

Facts and Skills

1. **Explain** what it means to precycle, reuse, and recycle.
2. **List** three ways to help make your environment positive.

Critical Thinking

3. **Summarize** Why is it important to reuse items?
4. LIFE SKILLS **Be a Health Advocate** You are designing a poster to communicate the benefits of recycling. What reasons to support recycling could you list on the poster?
5. LIFE SKILLS **Analyze What Influences Your Health** How does a positive environment influence your health?

CHAPTER 10 REVIEW

Use Vocabulary

conservation, *E47*
fossil fuels, *E48*
nonrenewable resource, *E48*
pollution, *E33*
positive environment, *E52*
reuse, *E51*

Choose the correct term from the list to complete each sentence.

1. When land, water, or air contain harmful substances, making them unhealthful to the environment and people, it is called __?__.
2. Coal, oil, and natural gas burned for energy are types of __?__.
3. An environment that promotes physical, mental and emotional, and family and social health is a __?__.
4. Something that cannot be replaced within a reasonable amount of time is a __?__.
5. The protection and careful use of the forests, rivers, minerals, and other natural resources is __?__.
6. When you use something again instead of throwing it away and buying a new one, you __?__ the item.

Review Concepts

Answer each question in complete sentences.

7. Why is a healthful environment important?
8. Describe two ways to reduce noise pollution.
9. How is untreated sewage harmful?
10. List three sources of renewable energy.
11. What do you do when you precycle?
12. Identify three sources of water pollution.
13. Why is it important to use renewable resources instead of nonrenewable resources?

Reading Comprehension

Answer each question in complete sentences.

The Clean Water Act limits the amount of pollutants that can be released into water. The sewage of most communities must pass through sewage treatment plants. These plants clean many harmful substances out of sewage. This makes it safer to release the sewage into rivers, lakes, or oceans.

14. What is the purpose of the Clean Water Act?
15. What do sewage treatment plants do?
16. What can you do to help limit water pollution?

Critical Thinking/Problem Solving

Answer each question in complete sentences.

Analyze Concepts

17. What is the connection between nonrenewable energy resources and pollution?

18. How does turning off lights when they aren't needed help reduce pollution?

19. List three types of pollution that you can help prevent. Describe what actions to take to prevent them.

20. How does recycling help to reduce waste?

21. How does the way people take care of their lawns affect water pollution?

22. Explain how our water supply might be different if the Safe Drinking Water Act did not exist.

Practice Life Skills

23. **Set Health Goals** Kiran has the habit of leaving the lights and TV on when he leaves a room. Describe a plan for changing this behavior.

24. **Make Responsible Decisions** You are shopping with your family for a new television set. You find two that are almost alike. One uses less energy and the other has a lower price. Which should you choose? Use the *Guidelines for Making Responsible Decisions*™ to help you decide.

Read Graphics

The chart shows worldwide use of fossil fuels between 1950 and 1998. Use it to answer questions 25–27.

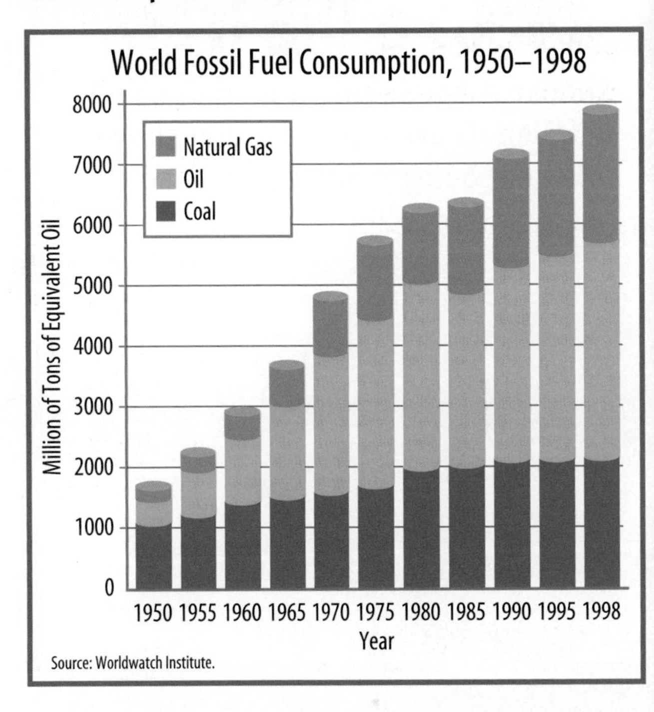

25. Of the three fossil fuels, which provided the largest share of energy in 1950? In 1998?

26. Between which years did the use of oil decline? Has the use of oil continued to decline?

27. About how many times more fossil fuel was being burned in 1998 than in 1950?

ACTIVITIES AND PROJECTS

Effective Communication

Draw a Diagram

Choose a household item that can be recycled. Draw a diagram that shows this item being manufactured, sold, used, and recycled.

Self-Directed Learning

Conduct an Interview

Choose a health career that interests you. With the help of your parents or guardian, contact someone with that career. Prepare your questions ahead of time. Meet with the person and ask your questions. Write down the answers. Report what you have learned to your class.

Critical Thinking and Problem Solving

Make a Picture Book

Brainstorm a list of items that could be reused rather than thrown out. Cut out pictures of these items from old magazines and newspapers and glue them to sheets of paper. Describe how each item could be reused under its picture. Collect the sheets and make a book.

Responsible Citizenship

Volunteer

Think of an activity you would like to volunteer for. With the help of your parents or guardian, find a group in your community that participates in this activity. Volunteer to spend a few hours helping the group.